D0866902

50/

SYMPOSIA OF THE
SOCIETY FOR EXPERIMENTAL BIOLOGY

NUMBER XII

Other Publications of the Company of Biologists

JOURNAL OF EXPERIMENTAL BIOLOGY
THE QUARTERLY JOURNAL OF MICROSCOPIC SCIENCE
JOURNAL OF EXPERIMENTAL MORPHOLOGY

SYMPOSIA

I NUCLEIC ACID

II GROWTH, DIFFERENTIATION AND MORPHOGENESIS

III SELECTIVE TOXICITY AND ANTIBIOTICS

IV PHYSIOLOGICAL MECHANISMS IN ANIMAL BEHAVIOUR

V FIXATION OF CARBON DIOXIDE

VI STRUCTURAL ASPECTS OF CELL PHYSIOLOGY

VII EVOLUTION

VIII ACTIVE TRANSPORT AND SECRETION

IX FIBROUS PROTEINS AND THEIR BIOLOGICAL SIGNIFICANCE

X MITOCHONDRIA AND OTHER CYTOPLASMIC INCLUSIONS

XI BIOLOGICAL ACTION OF GROWTH SUBSTANCES

The Journal of Experimental Botany is
published by the Oxford University Press
for the Society of Experimental Biology

SYMPOSIA OF THE
SOCIETY FOR EXPERIMENTAL BIOLOGY

NUMBER XII

THE
BIOLOGICAL REPLICATION
OF MACROMOLECULES

Published for the Company of Biologists
on behalf of the Society for Experimental Biology

CAMBRIDGE: AT THE UNIVERSITY PRESS

1958

PUBLISHED BY

THE SYNDICS OF THE CAMBRIDGE UNIVERSITY PRESS

Bentley House, 200 Euston Road, London, N.W. 1
American Branch: 32 East 57th Street, New York 22, N.Y.

Printed in Great Britain at the University Press, Cambridge
(Brooke Crutchley, University Printer)

CONTENTS

Self-reproduction and All That *p.* I
 by G. PONTECORVO

Fractionation of Deoxyribonucleic Acids and Reproduction of T2
 Bacteriophage 6
 by G. L. BROWN *and* A. V. BROWN

Studies of Deoxyribonucleic Acids with the Aid of Anion
 Exchangers 31
 by AARON BENDICH, HERBERT B. PAHL, HERBERT S. ROSEN-
 KRANZ *and* MORTON ROSOFF

Size Limitations Governing the Incorporation of Genetic Material
 in the Bacterial Transformations and Other Non-Reciprocal
 Recombinations 49
 by ROLLIN D. HOTCHKISS

Interspecific Reactions in Bacterial Transformation 60
 by PIERRE SCHAEFFER

Genetic and Physical Determinations of Chromosomal Segments in
 Escherichia coli 75
 by FRANÇOIS JACOB *and* ÉLIE L. WOLLMAN

Fertility Factors in *Escherichia coli* 93
 by HELEN L. BERNSTEIN

Colicins and Colicinogenic Factors 104
 by PIERRE FREDERICQ

Replication of an Animal Virus 123
 by F. K. SANDERS, J. HUPPERT *and* J. M. HOSKINS

On Protein Synthesis 138
 by F. H. C. CRICK

Protein Synthesis as Part of the Problem of Biological Replication 164
 by J. L. SIMKIN *and* T. S. WORK

Formation of Amylase in the Pancreas 176
 by F. B. STRAUB

The Biosynthesis of Oligo- and Polysaccharides *p.* 185
   by M. STACEY

Possible Mechanisms by which Information is Conveyed to the
   Cell in Enzyme Induction 195
   by M. R. POLLOCK *and* J. MANDELSTAM

Processes Co-ordinating Intracellular Activity 205
   by ALFRED MARSHAK

Tissue Transplantation and Cellular Heredity 225
   by N. A. MITCHISON

L'Action Inhibitrice du Glyoxal sur les Macromolécules
   Biologiques 242
   by J. ANDRÉ THOMAS

SELF-REPRODUCTION AND ALL THAT

By G. PONTECORVO

Department of Genetics, University of Glasgow

A biological system is defined as self-reproducing if (a) it is necessary for its own replication, i.e. it is not made in the absence of pre-existing ones of the same kind, and (b) it can give origin to a 'mutated' system which in its turn is capable of self-reproduction in the mutated form. Clearly, self-reproduction, or genetic continuity as it is otherwise called, is not the exclusive prerogative of cells or multicellular organisms: it is shown also by certain subcellular systems. Of these the oldest known and best studied are the chromosomes and their parts; this paper will refer abundantly to them.

The problems set by self-reproduction of subcellular systems have remained unchanged over the last 30 years—hence, incidentally, the somewhat facetious title of this paper—i.e. since they were first stated in a clear manner by Muller (1929) at the 1926 Ithaca symposium on 'The Gene'. However, we can now look at these problems with a wider background of information. Some of the reasons which were suggested then for rejecting certain ways of attack may perhaps be not so compelling today.

One of the developments which make this reappraisal possible is a better understanding of where the specificity of certain self-reproducing subcellular systems could lie, and how this specificity (or 'information' as we call it today) is conveyed from cell to cell. In fact, the realization that there are many ways in which this information can be transmitted is one of the most significant advances in biology in this century. It has led, among other things, to a unified view of cellular heredity, heredity in sexual reproduction, and infection (Darlington, 1944; Medawar, 1947; Lederberg & Lederberg, 1956).

On general grounds, one can conceive of two kinds of self-reproducing subcellular systems: on the one hand those which do not require a microscopic or megamolecular structural basis and which involve chains or cycles of reactions which are not spatially organized. The theory of such systems has been developed mainly by Hinshelwood (1947). Delbrück (1949) and Pollock (1953) have given models of how any one of them could determine alternative information. So far as I am aware, however, no convincing example of this type of system has yet come to light. The general feeling is that they should exist. The techniques for showing that they do, however, are barely existent.

On the other hand one can deduce on general grounds the existence, and in fact find examples in nature, of self-reproducing systems which have a structural basis of microscopic or at least megamolecular size. I shall deal here only with this kind of system.

THE DIVISIBILITY OF GENETIC INFORMATION

The naïve but useful early picture of the chromosome as a string with beads meant that a chromosome was made up of, say, 1000 active specific groups, the genes, held in a line by non-specific material. Each reciprocal exchange taking place at meiosis between the two members of a pair of homologous chromosomes was supposed to result from breakage of an intergenic bond at corresponding positions in each homologue followed by reunion in a crossing over-like fashion.

The picture which is taking shape now is more in line with Schroedinger's (1945) 'aperiodic' crystal. The evidence both from electron-microscope cytology and from genetic analysis suggests that there is no need for assuming lengthwise differentiation of the ultimate strand of the chromosome of a nature different from that, e.g. of a polypeptide or polynucleotide chain or a bundle of such chains. The 'beads' are epiphenomena; they are not basic structural singularities but they result from interactions between parts of the basic structure.

The evidence for this from electron-microscopy is due mainly to Ris (1957): the ultimate structure found within visibly differentiated sections of a chromosome—for instance the bands in a salivary gland chromosome or the chromomeres in leptotene chromosomes—is the same as that found anywhere else along the chromosome, and it is based on elementary fibrils of nucleoprotein about 200 Å. thick. The chromomeres and the bands are only regions of greater packing and twisting.

The evidence from genetic analysis is that a gene is separated from a neighbouring gene by linkages which are not distinguishable from those which separate the linearly arranged parts of the gene (Pontecorvo, 1952; 1956). In other words, recombination takes place both between genes and between parts of a gene.

All this does not exclude the possibility that there are along the chromosome, at intervals, bonds of a different nature. What we can say already is that these bonds of a special type (e.g. Mazia, 1954), if they exist at all, need not be those which delimit functional units. To take an old analogy, if we represent the building blocks of the elementary chromosome fibril as letters of the alphabet, the genes can be compared to the words of a sentence in which there is no difference between the spacing of letters and that of words. If there is punctuation, which interrupts occasionally the continuity of the sequence of letters, it would affect the meaning of the writing but usually only at a level higher than that of the individual words.

The chromosome, thus, appears as a structure the lengthwise specific activities of which (the genes) are in first approximation discrete, but not because the structure itself has corresponding discontinuities.

The difficulties, pointed out by Muller in 1926, for conceiving the gene as compound (i.e. made up of a linear array of *identical* units) are vindicated. The gene is not compound but complex, i.e. made up of a number of different elements integrated in activity. These elements one can identify, count and locate in linear sequence, thanks to the fact, known only since 1950, that recombination can occur between them as well as between genes. I have called these sub-gene elements *mutational sites* (Pontecorvo, 1952) and they correspond approximately to 'recons' in Benzer's (1957) terminology, i.e. to 'the smallest element in the one-dimensional array which is interchangeable (but not divisible) by genetic recombination'.

In *Aspergillus* the smallest recombination fractions measured between mutational sites of a gene (*ad8*, Pritchard, 1955) are of the order of 10^{-5} and the largest of 10^{-3}. This gives a *minimum* estimate of the number of sites per gene (or 'recons' per 'cistron') of the order of 100. In *Drosophila* a similar estimate for the *w* gene is about 60 (Pontecorvo & Roper, 1956).

In bacteriophage it is known that the genetic information is carried by DNA. With certain reasonable assumptions Benzer (1955, 1957) was therefore able to carry out the elegant translation of data of the kind mentioned above into lengths of DNA. The startling result is that in bacteriophage the ultimate unit of recombination and of mutation could well be a single nucleotide pair, while a gene (or 'cistron') would be based on about 1000 nucleotide pairs.

Following Benzer's lead we have attempted the same kind of translation for *Aspergillus* and *Drosophila* (Pontecorvo & Roper, 1956). In spite of the fantastic assumptions which one has to make in these cases, the results are surprisingly similar to those with bacteriophage: the smallest recombination fractions *so far measured* are equivalent to about 8 nucleotide pairs in *Aspergillus* and 200 in *Drosophila*, and the size of a gene ('cistron') of the order of 100–10,000 nucleotides in both.

Clearly we are still very far from identifying individual sites with individual nucleotide pairs in a polynucleotide sequence. In spite of the difficulties and the pitfalls, however, this seems to be one of the promising approaches. It is the more promising since Ingram (1957) has produced the startling news that haemoglobin in sickle cell anaemia seems to differ from normal haemoglobin by a single amino acid residue out of about 300. This difference is known to be determined by a single mutant gene.

CONCLUSION

From what I said above, a gene mutation may well consist in a change of a few nucleotide pairs, or even of only one. The possibility of translating a polynucleotide code in the chromosome into a polypeptide code in a gene-determined protein is no longer inconceivable. It will require the combination of genetical analysis as refined as that of Benzer (1957) for rII in phage, or of Pritchard (1955) for *ad8* in *Aspergillus*, with the analysis of a protein as refined as that of insulin and of a nucleic acid more refined than anything done so far.

The conclusion is that in the case of chromosomes and bacteriophages, the ultimate unit of mutation and of recombination may well be a single nucleotide pair. But this leaves us in a difficulty as to what is the ultimate unit of replication. Clearly a single nucleotide pair does not fit the definition of a self-replicating system because it can be synthesised *in vitro*. The significant feature is the *arrangement* of a particular nucleotide in a definite sequence of nucleotides. How long has the sequence to be in order that the cell cannot make it without a pre-existing one?

This paradox is precisely what we need as a conclusion. It suggests that the definition of self-reproduction given at the beginning of this paper is inadequate: crystallization, for instance, fits that definition.

To make that definition more precise, we need the idea that a collection of building blocks (say, of the four main types of nucleotide), which could arrange themselves into any one of all the possible sequences, becomes arranged into only one sequence *because of the pre-existence of such a unique sequence.*

REFERENCES

BENZER, S. (1955). *Proc. Nat. Acad. Sci., Wash.* **41**, 344.
BENZER, S. (1957). In *The Chemical Basis of Heredity*. Ed. McElroy, W. D. & Glass, B. Baltimore: Johns Hopkins Press.
DARLINGTON, C. D. (1944). *Nature, Lond.* **154**, 164.
DELBRÜCK, M. (1949). Discussion of paper by Sonneborn & Beale, pp. 33–4, in *Unités Biologiques douées de continuité génétique*, Paris, Centre Nat. Recherche Scient.
HINSHELWOOD, C. (1947). *The Chemical Kinetics of the Bacterial Cell*. Oxford: Clarendon Press.
INGRAM, V. M. (1957). *Nature, Lond.* **180**, 326.
LEDERBERG, J. & LEDERBERG, E. (1956). In *Cellular Mechanisms in Differentiation and Growth*. Princeton University Press.

MAZIA, D. (1954). *Proc. Nat. Acad. Sci., Wash.* **40**, 521.
MEDAWAR, P. B. (1947). *Biol. Rev.* **22**, 360.
MULLER, H. J. (1929). *Proc. Int. Congr. Plant. Sci.* **1**, 897.
POLLOCK, M. R. (1953). *Soc. Gen. Microbiol., 3rd Symp.*, p. 150.
PONTECORVO, G. (1952). *Adv. Enzym.* **13**, 121.
PONTECORVO, G. (1956). *Cold Spr. Harb. Symp. Quant. Biol.* **21**, 171.
PONTECORVO, G. & ROPER, J. A. (1956). *Nature, Lond.* **178**, 83.
PRITCHARD, R. H. (1955). *Heredity*, **9**, 343.
RIS, H. (1957). In *The Chemical Basis of Heredity.* Ed. McElroy, W. D. & Glass, B. Baltimore: Johns Hopkins Press.
SCHROEDINGER, E. (1945). *What is Life?* Cambridge University Press.

FRACTIONATION OF DEOXYRIBONUCLEIC ACIDS AND REPRODUCTION OF T2 BACTERIOPHAGE

By G. L. BROWN and A. V. BROWN

Medical Research Council Biophysics Unit,
Wheatstone Laboratory, King's College, London, W.C. 2

INTRODUCTION

The discovery of the role of deoxyribonucleic acids in the transformation of pneumococcal types by Avery, MacLeod & McCarty (1944) and in the reproduction of T2 bacteriophage by Hershey & Chase (1952) has shown that hereditary determinants of pneumococcus cells and of T2 bacteriophage can be transmitted from one bacterial cell to another in the form of deoxyribonucleic acid molecules. The evidence that deoxyribonucleic acids, where they are present, carry hereditary determinants in other biological systems is circumstantial in nature as yet and has been recently summarized by Hotchkiss (1955).

The analyses of the contents of purine and pyrimidine bases in deoxyribonucleic acids prepared from many different species of organism, summarized in a recent review by Chargaff (1955), suggest that the genetic specificity of deoxyribonucleic acid molecules might reside in the content and arrangement of these components. It was found that, whereas the ratios of purines to pyrimidines, adenine to thymine, and guanine to cytosine, were approximately equal to 1 in all preparations, the ratios of adenine + thymine to guanine + cytosine varied from 0·4 to 2 and were characteristic of the species. Wilkins (1956) has recently described evidence obtained by him and a number of collaborators using X-ray diffraction techniques, showing that all deoxyribonucleic acids so far examined contain molecules of the same overall configuration, a modified form of the double-stranded helical model put forward by Crick & Watson (1954) on the basis of the analytical data described above and of the X-ray data of Wilkins, Stokes & Wilson (1953) and of Franklin & Gosling (1953). Whether all the molecules in any particular preparation have this configuration cannot be established at present, but from these results it appears unlikely that genetic specificity in the molecules is due to differences in configuration. Thus the most probable way in which a deoxyribonucleic acid molecule can carry genetic information is by the arrangement of purines and pyrimidines in a specific sequence which can be translated by the synthetic processes of, say, a bacterial cell into a specific protein. This hypothesis, to which no reasonable alternative has been proposed, leads to the consideration of two problems. The first one is the determination of the way of governing the arrangement of twenty amino acids in a polypeptide chain by the arrangement of four bases in a linear polynucleotide chain. Theoretical solutions of this problem, which has been called the coding problem, have been put forward and discussed by Gamow (1954), Gamow, Rich & Yčas (1955), Gamow & Yčas (1955) and Crick, Griffith & Orgel (1957) but experimental evidence is not yet available to enable a choice to be made or an alternative solution to be proposed. The second problem is the determination of the way in which deoxyribonucleic acid molecules control the synthesis of a specific protein. Brachet (1941) and Caspersson (1941) first put forward the idea that ribonucleic acid is involved in protein synthesis, and the biological role of ribonucleic acid has been reviewed by Brachet (1955). The recent observation of Hoagland, Zamecnik & Stephenson (1957) that a complex between amino acid and soluble ribonucleic acid is an intermediate in the incorporation of the amino acid into a peptide linkage in a microsome system, is particularly suggestive of the part played by ribonucleic acid in protein synthesis. The role of deoxyribonucleic acid in protein synthesis, however, is not at all clear. The simple idea that deoxyribonucleic acids control the synthesis of a specific ribonucleic acid which then directs the synthesis of the corresponding proteins seems to be a doubtful one in view of the work of Brachet & Chantrenne (1956). The direct participation of deoxyribonucleic acids in stimulating amino acid incorporation into the proteins of isolated calf thymus nuclei has been observed by Mirsky, Osawa & Allfrey (1956), but, as the deoxyribonucleic acids can be replaced by various degradation products, the specificity of the processes involved may be questioned. Gale & Folkes (1955 b) found that deoxyribonucleic acids from *Staphylococcus aureus* stimulated the increase of catalase and β-galactosidase activities in disrupted cell preparations of the same organism depleted of nucleic acids.

From these and many related studies it is evident that there are a number of difficulties in studying the effects of deoxyribonucleic acids on protein synthesis. An apparent stimulation of protein synthesis may be unspecific and therefore unrelated to the genetic determinants carried by the deoxyribonucleic acid molecules. Incorporation of labelled amino acids does not necessarily imply a net protein synthesis and may be due to exchange reactions as shown by Gale & Folkes (1955 b). Another serious difficulty is that deoxyribonucleic acid preparations are complex mixtures of different molecular species which may have different functions, as will be seen from the fractionation work to be described. To overcome these difficulties, an ideal model system for the study of the translation of a specific nucleotide sequence carried by a deoxyribonucleic acid molecule into a specific protein would have to meet the following requirements. From an organism with characters susceptible to genetic analysis a homogeneous deoxyribonucleic acid preparation must be obtainable in which all the molecules contain the genetic determinants for a known protein which can be readily detected and isolated for structural studies. The other requirement is a subcellular fraction of a suitable organism which retains the ability to synthesize protein but cannot form the specified protein until induced to do so by the presence of the deoxyribonucleic acid which is normally not present in the second organism. This model system would have seemed impossible to achieve a short time ago, but developments in apparently unrelated fields over the past few years have now made it possible to set up such a system. The development of a method of fractionating deoxyribonucleic acids which has enabled us to isolate a molecular species satisfying these requirements, and the preparation of a suitable subcellular fraction, are described in this contribution together with some preliminary results obtained with the system and a discussion of related work.

HETEROGENEITY OF DEOXYRIBONUCLEIC ACID PREPARATIONS

The first observation of biochemical heterogeneity of deoxyribonucleic acid preparations was made by Bendich (1952) who observed a heterogeneous metabolism in liver deoxyribonucleic acid fractions separated by solubility in dilute salt solutions and detected by differential labelling with ^{14}C in the bases. This was followed by the development of two methods of fractionation of deoxyribonucleic acids using the same principle of fractional dissociation from basic proteins in two different ways. Chargaff, Crampton & Lipshitz (1953) demonstrated that, by extracting nucleoprotein gels prepared by treatment with a protein-denaturing agent with increasing concentrations of sodium chloride, fractions with different contents of bases were obtained and the ratios of the bases varied in a regular way. Similar results were found by Brown & Watson (1953), using a chromatographic procedure with a column of histone immobilized on celite which was fractionated by gradient elution with sodium chloride solutions. Another chromatographic procedure, using anionic exchangers prepared from cellulose according to the methods of Peterson & Sober (1956), has been devised by Bendich, Fresco, Rosenkranz & Beiser (1955) which gives fractions of varying molecular weight and with irregular variations in the contents of bases in the fractions. The heterogeneity of deoxyribonucleic acids has also been demonstrated by Lucy & Butler (1954) and by Lerman (1955), using dissociation from proteins, and by Shooter & Butler (1955), who observed a distribution of sedimentation constants in dilute solutions in the ultra-centrifuge. All these observations show the impossibility at present of separating from most types of deoxyribonucleic acid preparation a fraction consisting of a single molecular species unless special cases are discovered. Fortunately T2 bacteriophage DNA is such a special case, as will be seen later.

The original method of Brown & Watson (1953) has been developed along the following lines. In the early work, histone, prepared from water-extracted nucleohistone by extraction with dilute hydrochloric acid, was found to be strongly adsorbed to kieselguhr on heating a mixture of them to 100° C., giving a material suitable for chromatography of deoxyribonucleic acids. This material had two disadvantages; the histone preparation was not homogeneous, and small amounts of histone were being eluted with the fractions of deoxyribonucleic acids and complicating the resulting chromatographic analysis. The latter defect was overcome by coupling the histone by diazo linkages from tyrosine and histidine groups in the proteins to p-amino benzyl cellulose by a method previously used by Campbell, Leuscher & Lerman (1951) to couple serum albumin to cellulose. The first difficulty was overcome by developing a method of fractionating histone and using one of the fractions. To fractionate histones, columns of phosphorylated cellulose, prepared according to Hoffpauir & Guthrie (1950) were used with pH and salt-gradient elution. The resulting elution pattern of histones, obtained from calf-thymus nucleohistone, prepared by a modification of the method of Crampton, Lipshitz & Chargaff (1954a), by extraction with 0·2N-HCl is shown in Fig. 1. Studies were made of the fractionating properties of the two most abundant histone fractions C and D and it was found that the use of D, the most basic of the two, gave the largest range of differences in base ratios of the fractions obtained from calf-thymus deoxyribonucleic acids. Accordingly, fraction D has been used in all the fractionation work described here.

Other basic proteins tested for fractionating purposes have either been found to give small variations in base ratios in fractions of calf-thymus deoxyribonucleic acids or else have been difficult to couple to cellulose owing to their low contents of histidine and tyrosine. It must be mentioned here that Lipshitz & Chargaff (1956), using a batch extraction method from nucleoprotein gels, have shown that a wider range of base ratios is found in fractions of wheat-germ deoxyribonucleic acids prepared by dissociation from the basic proteins with which they are associated in their native state than from those obtained by dissociation from artificial complexes with other basic proteins. Unfortunately, it has not yet been found possible to prepare purified histones from many sources. Indeed, some deoxyribonucleic acids do not appear to be associated with basic proteins in their native state. The use of specific histones to fractionate their related deoxyribonucleic acids cannot then be applied in most cases.

An argument against the specificity of histones in this connexion is provided by the finding that a fractionation with respect to base ratio is obtained only with histones that have been denatured. This has been found

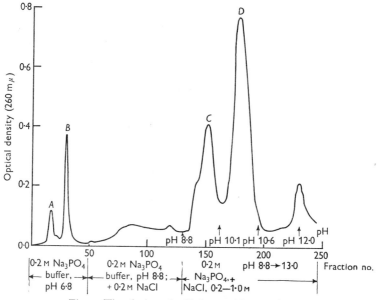

Fig. 1. The elution of calf-thymus histones from a column of phosphorylated cellulose.

by Crampton, Lipshitz & Chargaff (1954b) and by Lucy & Butler (1954) using batch methods and by the present authors using column methods. With fraction D of calf-thymus histone, it has been found that the amount bound to a particular batch of p-amino benzyl cellulose increased by 300 % after denaturing by heating to 100° C. for 15 min. Calf-thymus deoxyribonucleic acids were found to elute from columns of the denatured histone-cellulose complex over a wide range of molarities of the eluting sodium chloride solutions, as shown in Fig. 2, giving fractions with varying base ratios as shown in Table 1. Using columns of undenatured histone coupled to cellulose, however, the deoxyribonucleic acids eluted over a smaller range of sodium chloride concentrations with no appreciable variations in the base ratios of the fractions. These results imply that the polypeptide chain of the native histone molecule is folded so that some of the histidine and tyrosine groups are masked from the coupling reaction, and that the unfolded chain is necessary for the fractional dissociation of deoxyribonucleic acids from their complexes with basic proteins. The method now used by us to prepare the column material for the fractionation of deoxyribonucleic acids is as follows. Calf-thymus nucleohistone is prepared by the method of Crampton et al. (1954a) using a 3 hr. extraction with water

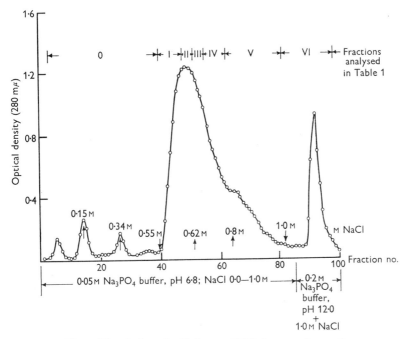

Fig. 2. The elution of calf-thymus DNA from a column of histone fraction *D* coupled to cellulose.

instead of 15 hr. The histones are extracted from this preparation with 0·2N-HCl, adsorbed on a phosphorylated cellulose column at pH 6·0, the column washed with 0·1M phosphate buffer at this pH and the basic components eluted from the column by a solution with continuously increasing pH and sodium chloride concentration between pH 9·0–13·0 and 0·2M–1·0M respectively. Component *D*, which elutes from the column between pH 10·1 and pH 10·4, is dialysed against water and denatured by heating to 100° C. for 15 min. After cooling it is reacted with diazobenzyl cellulose at 3° C. and any free diazo groups left are masked by reaction with β-naphthol. This material, which is exhaustively washed before use, absorbs from 1 to 3 mg./g. depending on the batch.

The procedure adopted for fractionation of deoxyribonucleic acids on cellulose-histone columns is to dissolve samples in aqueous sodium chloride solutions of low molarity and adsorb them on columns containing five times the amount of cellulose-histone necessary to adsorb all the material in the sample. The deoxyribonucleic acids are then eluted slowly from the column by a solution of buffered sodium chloride, the concentration of which is increased continuously. The variation of sodium chloride concentration is produced by using cylindrical reservoirs connected by siphons in which the type of concentration gradient applied to the column, linear, concave upwards or convex upwards, can be adjusted to suit the problem by using reservoirs of suitable diameters. The elution pattern of a calf-thymus deoxyribonucleic acid, prepared according to Kay, Simmons & Dounce (1952), obtained from such a column using a linear gradient is

Table 1. *Base analyses of fractions of calf-thymus DNA obtained by elution with sodium chloride solutions from histone-cellulose*

Fraction	0	I	II	III	IV	V	VI	Unfract.
Salt conc. (M)	0·0–0·5	0·55–0·6	0·6–0·65	0·65–0·7	0·7–0·8	0·8–1·0	1·0 [pH 12·0]	—
Adenine	31·6	23·9	25·8	28·6	30·9	33·8	28·5	28·8
Thymine	20·2	24·3	27·4	29·3	30·2	32·4	28·7	28·4
Guanine	28·7	26·2	23·1	19·7	19·3	16·2	21·2	21·5
Cytosine	19·5	25·6	23·7	22·4	19·6	17·6	21·6	21·3
$\frac{A+T}{G+C}$	1·07	0·93	1·14	1·37	1·57	1·96	1·34	1·34
$\frac{A+G}{T+C}$	1·51	1·01	0·95	0·94	1·01	1·00	0·98	1·01
ϵ_p^{260}*	8200	6420	6650	6540	6470	6680	7240	6580

* ϵ_p^{260} = optical density at 260 mμ of a DNA solution containing 1 mole of phosphorus per litre.

shown in Fig. 2 and the characteristics of the various fractions are summarized in Table 1. The fractions eluting between 0 and 0·4M-NaCl appear to be polynucleotides of low molecular weight, as their values for ϵ_p are high and the amounts eluting in these fractions increase during the course of deoxyribonuclease digestion. The main bulk of the material is eluted from 0·55M-NaCl to 1M-NaCl and to judge from the ϵ_p^{260} data has retained its native properties. On analysis of different parts of this peak for purine and pyrimidine bases, however, the contents of these vary considerably but in a regular way. The ratios of adenine + thymine to guanine + cytosine increase from approximately 0·9 to nearly 2·0 with increasing sodium chloride concentration but the ratios of purines to pyrimidines remain approximately equal to 1. Fractions taken from different parts of this peak were found by Wilson & Wilkins (1955) to give the X-ray diffraction pattern of a double helix, showing that all contained molecules with this configuration. On re-running fractions taken from different parts of the main peak an interesting and useful effect was observed. The fractions taken from near the leading edge of the main peak gave sharper peaks on re-running than those from near the rear edge. A study of this effect has shown that a displacement effect, produced by the more strongly binding molecular species displacing the more weakly binding species from the binding sites during the elution, increases the resolution of the fractionation of the material eluting at the lower salt concentrations. This effect can be accentuated by using long columns and salt gradients which are concave upwards. On further elution with 1·0M-NaCl, 0·1M phosphate buffer pH 12·0 another fraction is obtained which has the base content of the original material. The amount of material in this fraction increases with the time that the deoxyribonucleic acid is left adsorbed to the histone-cellulose before elution, and seems to be a column artifact. To minimize this high pH fraction, the sample is put on the column at 0·4M-NaCl and the material eluting below this concentration washed through the column and refractionated between 0 and 0·4M-NaCl separately. The main bulk of the deoxyribonucleic acid is then eluted by a gradient of sodium chloride concentration between 0·4M and 1·0M-NaCl and the high pH fraction by the pH 12·0 buffer.

The results obtained with calf-thymus deoxyribonucleic acids have been described in detail because the same overall features of the elution pattern have been found with deoxyribonucleic acids prepared from avian tubercle bacillus, *Escherichia coli*, human lymphocytes (in collaboration with L. D. Hamilton), pneumococci (in collaboration with H. Ephrussi-Taylor) and *Arbacia lixula*. In each case the elution pattern below 0·4M-NaCl varies with the preparation but never accounts for more than 5% of the total phosphorus of the sample. The concentration of sodium chloride at which the bulk of the deoxyribonucleic acids starts to elute is characteristic of the source and corresponds approximately to the overall base ratio of the specimen. The amount of material eluting at pH 12·0 is a function of the time the material is left adsorbed to the column and increases from 2% at 10 hr. to about 10% at 60 hr., the amount of material eluting at neutral pH's decreasing accordingly.

The difference in the behaviour of deoxyribonucleic acids and ribonucleic acids on the cellulose-histone columns is shown in Fig. 3, which is the elution pattern of a mixture of *Escherichia coli*-deoxyribonucleic acid prepared by the method of Gandelman, Zamenhof & Chargaff (1952), and of a highly polymerized *E. coli*-ribonucleic acid prepared by a modification of the method of Crestfield, Smith & Allen (1955). Most ribonucleic acids have been found to elute from the columns at about 0·3M-NaCl. In Table 2 are shown the variations of base contents in different sections of the main elution peak of *E. coli*-deoxyribonucleic acid, which can be seen to be much smaller than in the case of calf-thymus deoxyribonucleic acid. This may be due to there being a narrower range of variation of base contents in bacterial deoxyribonucleic acids. A comparison, however, of the elution patterns of avian tubercle- and of *E. coli*-deoxyribonucleic acids indicates that the resolving power of the cellulose-histone column with

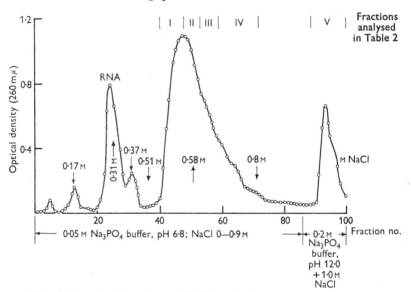

Fig. 3. The elution of an 8:1 mixture of DNA and RNA, prepared from *Escherichia coli*, from a column of histone fraction *D* coupled to cellulose.

Table 2. *Base analyses of fractions of* Escherichia coli *DNA obtained by elution from histone-cellulose columns with sodium chloride solutions*

Fractions	I	II	III	IV	V	Unfract.
Salt conc. (M)	0·51–0·58	0·58–0·63	0·63–0·70	0·70–0·80	1·0 [pH 12·0]	—
Adenine	23·5	23·9	25·4	27·6	25·6	25·5
Thymine	22·7	24·3	25·8	26·8	25·0	24·4
Guanine	25·7	26·0	23·6	22·6	25·4	24·5
Cytosine	28·1	25·8	25·2	23·0	24·0	25·6
$\dfrac{A+T}{G+C}$	0·86	0·93	1·05	1·19	1·02	0·99
$\dfrac{A+G}{T+C}$	0·93	0·99	0·96	1·08	1·04	1·00

respect to separation of molecular species with different ratios of adenine + thymine to guanine + cytosine is negligible for species with this ratio less than 1.

In collaboration with H. Ephrussi-Taylor, a study of the fractionation of *Pneumococcus*-deoxyribonucleic acids with transforming activities has been carried out, mostly with columns of fractionated histone adsorbed on kieselguhr. It was found that all the transforming activities were eluted in the main fraction between 0·69M and 0·95M-NaCl, and that material eluting below 0·4M-NaCl was inactive. On following the elution of streptomycin-resistance transforming activity, it was found that this activity was not eluted parallel to the elution of the deoxyribonucleic acid but was concentrated near the leading edge of the main elution peak as shown in Fig. 4. Fractions were obtained in this way with 5 times the

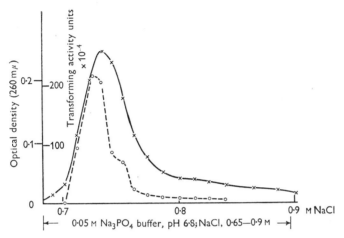

Fig. 4. The elution of pneumococcus-DNA with streptomycin-resistance transforming activity from a column of histone fraction *D* adsorbed on kieselguhr. —×—×—×—, Optical density at 260 mμ versus molarity of sodium chloride in fractions; - - O - - O - - -, streptomycin-resistance transforming activity versus molarity of sodium chloride in fractions.

specific activity of the starting material. On repeating this experiment with a preparation containing streptomycin-resistance and canavanine-resistance transforming activities both were found to be eluted in parallel with each other and no separation of these properties was observed.

The results summarized here, together with those obtained by the methods of Chargaff and his collaborators, show that the binding of deoxyribonucleic acid molecules to denatured histones increases with the ratio of adenine + thymine to guanine + cytosine. A satisfactory explanation for this effect has not yet been found. The explanation originally put forward by Brown & Watson (1953), based on some work by Cavalieri (1952), was that hydrogen-bonding between the 2-amino group of guanine and a neighbouring phosphate group could reduce the acidic properties of the latter, and therefore the binding of the deoxyribonucleic acid molecule would be reduced as their guanine content increased, as observed experimentally. This explanation is not compatible with the configuration of the molecules as found by X-ray diffraction methods and must be discarded. The structural basis of variation of binding strength with base composition remains obscure, then, but provides a useful method for a very partial separation of the highly complex mixtures of molecular species that compose the usual preparations of deoxyribonucleic acids. It would be rash, however, to expect such a crude method, even in combination with the methods of Bendich and his collaborators, to separate a homogeneous and distinct molecular species. Even a fraction with a well-defined base content could contain a large number of molecular species with different nucleotide sequences. T2 bacteriophage-deoxyribonucleic acid has some peculiarities which enable us to evade some of these difficulties.

THE FRACTIONATION OF T2 BACTERIOPHAGE-DEOXYRIBONUCLEIC ACID

Bacteriophages, especially the even-numbered T set, provide a source of deoxyribonucleic acid, the genetic role and function of which has been more intensively studied than in any other case. This makes fractionation studies on deoxyribonucleic acids prepared from them especially interesting.

By extraction of gels of T6r$^+$ deoxyribonucleic acid and histone with sodium chloride solutions, Crampton *et al.* (1954*a, b*) obtained fractions with ratios of adenine + thymine to guanine + 5-hydroxy-methylcytosine varying from 1·6 to 1·98. Bendich, Pahl & Beiser (1956) have studied the fractionation of T6r and T6r$^+$ deoxyribonucleic acids on a substituted cellulose ion-exchanger and have found large differences between them.

Using cellulose-histone columns, Brown & Martin (1955) found that the elution profile of T2r deoxyribonucleic acid was different from the usual one obtained with preparations from other sources as shown in Fig. 5. Less than 1 % of the material was eluted below 0·4M-NaCl and about 90 % began to elute in a double peak at 0·69M-NaCl. This double peak was accentuated by using columns up to 80 cm. long and a concentration gradient of sodium chloride which was concave upwards. By analysis of the material in each peak and in different parts of each peak, and by re-running material from the peaks, it was shown that they represented two components *A* and *B* with ratios of adenine + thymine to guanine + 5-hydroxy-methylcytosine of about 1·9 and 2·15 respectively, constituting about 40 % and 60 % of the total deoxyribonucleic acid.

Soon after this finding, Levinthal (1955) demonstrated that T2 bacteriophage-deoxyribonucleic acids contained a large molecule with a molecular weight of about 50 million and a number of smaller pieces with molecular weights of less than 10 million. This was shown by growing T2 phage in a medium heavily labelled with ^{32}P, purifying the labelled phage, and studying the molecular autoradiographs obtained when the phage particles and osmotically shocked phage were mixed with Ilford G-5 nuclear research photographic emulsion. In this emulsion, which is sensitive to electrons and registers each β-particle from the disintegration of a ^{32}P atom as a track of silver grains, a phage particle gives a star of tracks emanating from a point. The mean number of tracks per star is proportional to the mean number of ^{32}P atoms in each phage particle and to the time between drying the emulsion and developing it. If the phage particles are disrupted by osmotic shock by the method of Anderson (1953) before being

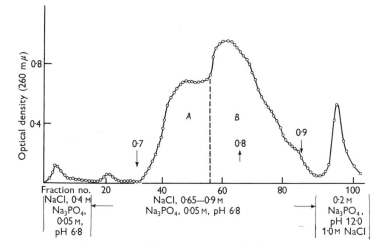

Fig. 5. The elution of T2r bacteriophage-DNA from a column of histone fraction D coupled to cellulose.

mixed with the emulsion, the same number of stars is observed at equivalent dilutions, but the stars now have only 40 % of the mean number of tracks per star of the phage particles. Levinthal & Thomas (1957a) have analysed this situation statistically and have shown that the large molecule in the osmotically shocked preparations is a unique piece and not due to the random breakdown or association of larger or smaller pieces respectively. The stars are not obtained after treatment with deoxyribonuclease but are unaffected by treatment with chymotrypsin. Levinthal & Thomas were unable to prove that the star-forming component was free of protein, and the possibility remained that the large piece of deoxyribonucleic acid from the phage was an association in a definite structure of smaller molecules with some protein, either the membrane protein or one of the non-sediment-able proteins that Hershey (1955) has found to be released from the phage head by osmotic shock. It seemed likely to us that one of our fractions might be the large piece of phage-deoxyribonucleic acid observed by Levinthal. One of us, in collaboration with N. W. Symonds, has investigated the fractionation on histone-cellulose of osmotically shocked T2r bacteriophage heavily labelled with ^{32}P following the elution of star-forming molecules by using the methods described by Levinthal & Thomas. The results are shown in Fig. 6. The elution of the total deoxyribonucleic acid was followed by measuring the radioactivity of samples of the fractions with a Geiger counter. It can be seen from Fig. 6 that the star-forming molecules are eluted at the beginning of the elution peak. This has a

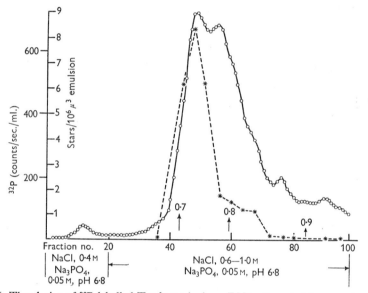

Fig. 6. The elution of ^{32}P-labelled T2r bacteriophage-DNA, disrupted by osmotic shock, from a histone-cellulose column, showing the distribution of ^{32}P star-forming molecules in the fractions compared with the total ^{32}P in the fractions. —O—O—O—, ^{32}P in fractions measured by counts/sec./ml.; - - *- - *- - *- -, ^{32}P stars per $10^6 \mu^3$ of G-5 emulsion obtained from samples of fractions.

different outline from that in the large-scale fractionation, as the elution is carried out as rapidly as possible to eliminate complications due to breakdown of the molecule when the ^{32}P atoms disintegrate. Also the loading of the column in this experiment is one-thousandth of that in the large-scale experiment. It seems almost certain, then, that our fraction A is identical with the high molecular weight fraction observed by Levinthal.

The possibility that the star-forming component might be a complex of small molecules with protein has been eliminated by repeating the fractionation with osmotically shocked phage particles labelled with ^{35}S. It was found in this case that all the radioactivity was washed through the column with 0·4M-NaCl.

The presence of glucose as a glycoside substituent in the 5-hydroxymethylcytosine of the deoxyribonucleic acids in the even-numbered T series of bacteriophage was first detected by Sinsheimer (1954). Later (1956), he found that the molar ratio of glucose to 5-hydroxy-methylcytosine in a strain of T2 was 0·77, but was 1·0 in a T4 strain, and that in enzyme digests of the latter there were no detectable 5-hydroxy-methylcytosine nucleotides unsubstituted with glucose. On examination of the ratios of glucose to 5-hydroxy-methylcytosine of our fractions A and B, in collaboration with S. Kosinski, we found the values 0·65 and 0·98 for A and B respectively. The observed properties of the components of T2 bacteriophage-deoxyribonucleic acid are summarized in Table 3.

Table 3. *Properties of the fractions of T2r bacteriophage-DNA*

Fractions	A	B	Unfract.
$\dfrac{\text{Adenine} + \text{thymine}}{\text{Guanine} + \text{5-OH-methylcytosine}}$	1·89	2·14	2·06
$\dfrac{\text{Glucose}}{\text{5-OH-methylcytosine}}$	0·65	0·98	0·84
Molecular wt. (from star tracks)	$\sim 50 \times 10^6$	$< 10 \times 10^6$	?

It appears, then, that the star-forming component lacks glucose in 35 % of its 5-hydroxy-methylcytosine groups. This is genetically interesting in view of the observations of Streisinger & Weigle (1956), who have studied the phenomena of partial exclusion in mixed infections with T2 and T4 bacteriophage, and its association with the amount of glucose-substituted 5-hydroxy-methylcytosine in the deoxyribonucleic acids of the phages. They found that, on simultaneous infection of *Escherichia coli* B cells with T2 and T4, the markers of T2 are partly excluded from the progeny but that practically all the progeny, including those with T2 markers, had the excluding power of T4 with respect to T2 phage. Two other properties were found to be associated with the excluding power, high efficiency of plating on a strain of *E. coli* K-12 on which T2 has a low efficiency of plating, and 100 % glucose substitution in the 5-hydroxy-methylcytosine of the deoxyribonucleic acid. Strains obtained from a T4 × T2 cross and then back-crossed to T2 nine times to make them isogenic with respect to T2 still retained the excluding power of T4, the high efficiency of plating on the K-12 strain and the 100 % glucose-substitution in the 5-hydroxy-methylcytosine. These strains were designated $\overline{\text{T2}}$, and the associated properties were called the bar properties.

From the fractionation data it is likely that the differences between T2 and $\overline{\text{T2}}$ are related to the differences in the amount of glucose-substitution in the star-forming component. The unusual inheritance pattern for the bar properties and the phenomena of partial exclusion in crosses of

$T_4 \times T_2$ and $T_2 \times \overline{T_2}$ may be due to differences in the rates of multiplication or to differences in stabilities towards breakdown by enzymes of star-forming molecules with different glucose contents. This explanation has especial point in view of the work of Levinthal (1956) which strongly indicates that the star-forming component may be considered as a phage chromosome. Levinthal has followed the star-forming molecules through several cycles of infection by the nuclear emulsion technique. He observed that the stars obtained from whole and shocked phage after a second cycle of infection had 20 % of the mean number of tracks per star of the original labelled phage and that this is maintained through further cycles of infection. Levinthal has interpreted this as showing that the star-forming molecule splits into two during replication. Levinthal & Thomas (1957b) have presented evidence that a host-range genetic factor originally associated with the radioactive parent in a cross of ^{32}P-labelled T2 phage and unlabelled phage was found to be associated with star-forming phage after further cycles of infection to eliminate phenotypic mixing. This observation suggests that the star-forming molecule carries host-range factors.

Although many of these observations have to be expanded and confirmed, they do show the possibility of fulfilling one of the requirements of the model system for studying the relation between deoxyribonucleic acids and protein synthesis. That is, the requirement for a homogeneous preparation of deoxyribonucleic acid carrying known genetic determinants. The star-forming component or our fraction A, which, if the work of Levinthal & Thomas (1957b) is confirmed, carries determinants for the phage protein that governs host range, will fill this requirement.

The completion of a model system, then, requires a subcellular bacterial system that can be induced to form the bacteriophage tail proteins by the bacteriophage-deoxyribonucleic acids. Such a system will now be described.

THE FORMATION OF BACTERIOPHAGE PROTEINS IN DISRUPTED PROTOPLASTS

The formation of bacterial protoplasts from *Bacillus megaterium* and related organisms by the controlled dissolution of the cell walls by lysozyme was first observed by Tomcsik & Guex-Holzer (1952) and Weibull (1953a). The preparation and properties of these protoplasts have recently been reviewed by Weibull (1956) and McQuillen (1956). Both Brenner & Stent (1955) and Salton & McQuillen (1955) have shown that multiplication of bacteriophage can take place in protoplasts of *B. megaterium* if the bacteria are infected with phage before their walls are digested with lysozyme.

Attempts by Brenner (1957) to produce phage in protoplasts by incubation with phage deoxyribonucleic acid have so far been unsuccessful.

Weibull (1953 b) has shown that if cells of B. *megaterium* are treated with lysozyme in the absence of a protoplast-stabilizing agent or if the amount of stabilizing agent in a suspension of protoplasts is reduced by dilution, then protoplast ghosts are formed. If the dilution is carried out in the presence of magnesium ions then the ghosts formed are more compact and less fragmented and retain more material than in its absence. Spiegelman (1957) has shown that the shocked protoplasts of B. *megaterium* retain ability to synthesize nucleic acids and proteins if suitable substrates are present.

Until recently, it was believed that *Escherichia coli* was resistant to lysozyme digestion, but Rapske (1956) has now shown that addition of versene after a short treatment with lysozyme at pH 8·1 removes the cell walls of this organism, leaving protoplasts stable in 0·5 M-sucrose. Two other methods of preparing *E. coli* protoplasts have been described by Lederberg (1956) and by Zinder & Arndt (1956). Mahler & Fraser (1956) have demonstrated that *E. coli* protoplasts can synthesize phage if the bacteria are infected before removal of the cell walls.

We have tested several bacterial systems, including ultrasonically disrupted *E. coli* cell preparations similar to those obtained from *Staphylococcus aureus* by Gale & Folkes (1955 a), for their ability to synthesize T2 phage proteins in the presence of the phage-deoxyribonucleic acids, but without success until the following system was examined.

Escherichia coli protoplasts were prepared by a modification of the method of Rapske (1956) and ghosted by a rapid one-in-eight dilution in 0·05 M potassium phosphate buffer pH 6·8 containing 0·003 M-MgSO$_4$ and 0·02 M-NaCl, in which the ghosts could be washed and kept for several hours without further loss of material. These preparations contain less than 0·1 % viable bacteria, which can be reduced to about 0·001 % by differential centrifugation. These protoplast ghosts were found to have lost the ability to incorporate ^{14}C-labelled amino acids in the presence of glucose so, presumably, the ghosting process had resulted in the loss of the ability to synthesize protein. It was clear, however, that during the ghosting process most of the soluble enzymes required for protein synthesis would have been lost and would have to be replaced at a higher concentration than the original concentration in the bacterial cells due to the dispersion in the incubation medium. The observation of the presence of amino acid-activating enzymes, similar to those discovered by Hoagland (1955) in rat-liver homogenates, in extracts of *E. coli* cells by De Moss & Novelli (1956) indicated that these in particular would be lost. Accordingly, an extract of *E. coli* containing the amino acid-activating enzymes was made by their method, passed through a cellulose-histone column at pH 8 to remove high molecular weight polynucleotides, and freeze-dried after dialysis. De Moss & Novelli also observed that only eight of the twenty amino acids commonly found in proteins are activated by the *E. coli* extracts. Cole, Coote & Work (1957) have recently demonstrated that cell extracts from various sources contain enzymes activating different ranges of amino acids. Therefore the *E. coli* extracts were supplemented by the amino acid-activating enzymes from rat liver prepared by the method of Hoagland, Keller & Zamecnik (1956), the precipitation at pH 5·1 being carried out after removal of polynucleotides by passage through a cellulose-histone column at pH 8·0.

The work of Hershey (1953) and of Volkin & Astrachan (1956) has indicated that the synthesis of a new ribonucleic acid might be an intermediate step in the induction by deoxyribonucleic acid of the synthesis of phage proteins in *E. coli* cells. It was decided, therefore, that the presence of ribonucleoside phosphorylase and its substrates, the four ribonucleoside diphosphates, guanosine, adenosine, uridine and cytidine diphosphates might be required as these would have been lost from the ghosts. As the specific activity with respect to this enzyme of our *E. coli* extracts was below 0·1 as measured by the method described by Grunberg-Manago, Ortiz & Ochoa (1956), the enzyme was prepared in higher concentrations from *Azotobacter vinelandii* by fractionating an extract with ammonium sulphate as described by them in the same paper, giving crude fractions with specific activities between 0·7 and 1.

To supply energy for protein synthesis in the protoplast ghost system, adenosine triphosphate and an ATP-generating system consisting of phosphoenolpyruvate and pyruvate kinase were added. All the twenty amino acids commonly occurring in proteins were added to the incubation mixtures.

The effects on the incorporation of ^{14}C-labelled L-isoleucine into the protein of the fortified *Escherichia coli* protoplast ghosts by these crude enzyme systems and co-factors are shown in Table 4. From these it can be seen that the amount of incorporation of L-isoleucine can be successively increased by all the additions listed. When the complete set of amino acids other than L-isoleucine is omitted, the incorporation of L-isoleucine decreases to about 30 % of that of the final system, suggesting that most of the incorporation may be due to a net synthesis of protein rather than to exchange. At this stage the fortified protoplast ghosts appeared to provide a good system for testing the effects of T2 bacteriophage-deoxyribonucleic acid and a possible induction of the synthesis of phage proteins.

T2 bacteriophage proteins can be detected and measured by two specific immunological methods using reactions with antiphage serum. The first method measures the fixation of complement by phage-antiphage serum complexes and has been developed for quantitatively measuring phage proteins by Lanni (1954). De Mars (1955) has developed another method of detecting and measuring phage proteins by measuring serum blocking power, that is the ability to decrease the neutralizing activity of antiphage serum. A study by Lanni & Lanni (1953) has shown that the complement-fixation method measures chiefly the head protein of T2 bacteriophage,

Table 4. *The effect of some enzymes and co-factors on the incorporation of* ^{14}C-*labelled* L-*isoleucine into the proteins of* Escherichia coli *protoplast ghosts*

Protoplast ghosts	Iso-leucine	ATP and generating system	Amino acid mixture	E. coli extract	Rat liver pH 5 enzymes	A. vine-landii extract	Ribo-nucleoside di-phosphates	Incorp. L-isoleucine (mμmoles/ mg./hr.)
+	+	−	−	−	−	−	−	0·03
+	+	+	−	−	−	−	−	0·74
+	+	+	+	−	−	−	−	1·8
+	+	+	+	+	−	−	−	5·3
+	+	+	+	+	+	−	−	5·9
+	+	+	+	+	+	+	−	6·4
+	+	+	+	+	+	+	+	7·6
+	+	+	−	+	+	+	+	2·2
−	+	+	+	+	+	+	+	0·01

The complete reaction mixture contained 10^{10} protoplast ghosts/ml.; 2 μmoles/ml., ATP; 10 μmoles/ml., phosphoenolpyruvate; 0·01 mg./ml., pyruvate kinase; 0·1 mg./ml., rat liver pH 5 enzymes; 0·1 mg./ml., polyribonucleotide phosphorylase, spec. activity 0·7, prepared from *Azotobacter vinelandii*; 1·0 mgm. each of the disphosphates of cytosine, uridine, guanosine and adenosine; 0·2 mg./ml., of ^{14}C-labelled L-isoleucine with a specific activity of 0·66 mc./m.mole; 0·2 mg./ml., with respect to the L-isomers of L-asparagine, L-aspartic acid, DL-alanine, L-arginine dihydrochloride, L-cysteine hydrochloride, DL-methionine, L-glutamic acid, glycine, L-cystine, L-histidine, L-leucine, L-lysine dihydrochloride, DL-phenylalanine, L-proline, DL-serine, DL-threonine, L-tyrosine, DL-tryptophan, L-valine.

whereas the serum blocking power measures the tail proteins responsible for the interaction of the phage with the bacterial surface and which therefore include the proteins determining the host range of a particular phage strain. Streisinger (1956) has shown, by a genetic study, a very close association of host range and serological specificity with neutralizing activity of antiphage sera. The method of complement fixation was found to be difficult to apply to the measurement of phage proteins in the fortified protoplast ghost system, since some of the necessary components produced lysis of the red blood cells needed to measure the degree of complement fixation.

The purification of the T2r phage-deoxyribonucleic acid to the required limits of purity was a serious difficulty. The serological methods for measuring the phage proteins are very sensitive and are capable of detecting as little as 10^8 phage equivalents of protein in 1 mg. of deoxyribonucleic acid, which corresponds to about 2×10^{-5} mg. protein or 0·002 % protein impurity. As the amounts of phage antigens synthesized in the protoplast ghosts were expected to be small, at least until the requirements of the system had been fully studied, the appearance of these amounts might have been masked if phage deoxyribonucleic acids with larger amounts of protein impurity than this were used. Two methods of preparing T2r phage-deoxyribonucleic acid with less than this amount of protein were devised, both using phage purified by the method of Herriott & Barlow (1952).

In method A, the phage preparation was osmotically shocked, and some of the phage ghosts removed by centrifugation. Most of the rest were adsorbed to bacterial membranes made by the method of Weidel (1951) and centrifuged down with them. The supernatant was then passed through a column of phosphorylated cellulose. This procedure was based on the finding of Puck & Sagik (1953) that the T phages are adsorbed and inactivated on the surface of cationic exchange resins. We have found phosphorylated cellulose very efficient at absorbing T2 bacteriophage and its ghosts. After passage through the column the material is precipitated in alcohol and stored at $-20°$ C.

In method B, the phage particles are suspended in 10 % NaCl and agitated with chloroform three times in a blendor and then a number of times with a 9:1 chloroform/octanol mixture, removing the denatured protein gel each time by centrifugation, until no more gel was observed at the interface of the aqueous and organic liquid phases. The deoxyribonucleic acid left in the aqueous phase was then absorbed on a column of cellulose-histone which was then washed thoroughly with 0·4M-NaCl to remove proteins. The material was then eluted from the column and precipitated with alcohol and stored at $-20°$ C.

Both preparations were tested for infectivity by plating with sensitive bacteria, and for the presence of phage antigens by the complement-fixation method and by the serum blocking-power method. All these tests were negative. Suitable concentrations of T2 phage were mixed with calf-thymus deoxyribonucleic acid and plated to test for any masking of infectivity by the presence of nucleic acids, but no inhibitory effect was observed. Therefore both preparations appeared to be completely free of infectious particles and to contain less than 10^8 phage equivalents of antigens per milligram.

The effects of these two types of preparation on the uptake of ^{14}C-labelled L-isoleucine in the complete protoplast ghost system after removal of the deoxyribonucleic acid in the protoplast ghosts by deoxyribonuclease are shown in Table 5. The treatment of the protoplast ghosts with deoxyribonuclease reduced the incorporation by about 30 %; this was more than completely restored by addition of 0·1 mg./ml. of preparation A but only partly by preparation *B*. A deoxyribonuclease digest of preparation *A* and a preparation of phage ghosts had little effect. These results showed an increase in incorporation of about 2·5 mμmoles of L-isoleucine per mg. protein induced by preparation *A*. If this was due to a net protein synthesis it would have corresponded to a synthesis of about 10^{10} phage equivalents of protein per mg. of total protein. We were encouraged by this result to try and detect the formation of phage-specific antigens.

Table 5. *The effects of T2 bacteriophage-DNA preparations on the incorporation of* ^{14}C-*labelled* L-*isoleucine into the proteins of* Escherichia coli *protoplast ghosts*

Bacterial preparation	Enzymes, amino acids, and co-factors	Bacteriophage preparation	Incorp. L-isoleucine (mμmoles/mg./hr.)
Protoplast ghosts (untreated)	Complete system*	—	7·3
Protoplast ghosts (digested with DNAase)	Complete system	—	5·1
Protoplast ghosts (digested with DNAase)	Complete system	0·1 mg./ml. T2 phage-DNA, prep. *A*	7·9
Protoplast ghosts (digested with DNAase)	Complete system	0·1 mg./ml. T2 phage-DNA, prep. *B*	6·2
Protoplast ghosts (digested with DNAase)	Complete system	0·1 mg./ml. T2 phage-DNA, prep. *A* (digested with DNase)	5·7
Protoplast ghosts (digested with DNAase)	Complete system	10^{12} per ml. T2 phage ghosts	4·9

* The complete system contains all the additions listed in Table 4.

To determine serum blocking power produced in a fortified protoplast ghost suspension a standard curve was obtained in phage equivalents by adding suitable dilutions of ultra-violet inactivated T2 phage particles to the complete protoplast system minus diphosphates to save expense. Otherwise the procedure of De Mars was followed. The results obtained by incubating the deoxyribonuclease-digested protoplast ghosts and the complete system of enzymes and co-factors together with the T2 bacteriophage-deoxyribonucleic acids are shown in Table 6. Samples from each incubation mixture were plated for bacterial counts and on sensitive bacterial cells for the presence of bacteriophage. In all cases the viable cell count, which was about 2×10^5 per mg., decreased slightly and no phage plaques were observed. These results provide clear evidence that the protoplast ghost system has formed measurable quantities of T2 phage-specific proteins closely related to the tail proteins under the influence of the phage-deoxyribonucleic acids.

At first we were inclined to attribute the differences between the efficiencies of the two types of deoxyribonucleic acid preparations in stimulating the formation of phage protein to damage caused by the violent agitation used in preparation B. On testing fractions of preparation A obtained by the methods described earlier, using cellulose-histone columns, for their ability to induce the formation of proteins with serum blocking power, we obtained no effect. It was also found that on simply absorbing the deoxyribonucleic acid on a cellulose-histone column at low salt concentrations and eluting it at high salt concentrations, the ability was lost. The

Table 6

Bacterial preparation	Enzymes, amino acids, and co-factors	Bacteriophage preparation	Incubation temperature (° C.)	Serum blocking power phage equiv. per mg. protein per hr.
Protoplast ghosts (digested with DNAase)	Complete system	—	37	None detectable
Protoplast ghosts (digested with DNAase)	Complete system	0·1 mg./ml. T2 phage-DNA, prep. A	0	None detectable
Protoplast ghosts (digested with DNAase)	Complete system	0·1 mg./ml. T2 phage-DNA, prep. A	37	$2·4 \times 10^9$
Protoplast ghosts (digested with DNAase)	Complete system without diphosphates	0·1 mg./ml. T2 phage-DNA, prep. A	37	None detectable
Protoplast ghosts (digested with DNAase)	Complete system	0·1 mg./ml. T2 phage-DNA, prep. A (digested with DNase)	37	None detectable
Protoplast ghosts (digested with DNase)	Complete system	0·1 mg./ml. T2 phage-DNA, prep. B	37	$\sim 2·0 \times 10^8$

possibility that such a brief absorption would damage the deoxyribonucleic acid seemed unlikely as the streptomycin-resistance transforming ability of a *Pneumococcus*-deoxyribonucleic acid was found to be unaffected by a much longer absorption under similar conditions.

Preparation A was then repeated with ^{35}S-labelled T2 phage and it was found that, although phage antigens were absent, the material contained 0·3 % protein. This protein was found to be unabsorbed by bacteria, not precipitated by antiphage serum in the presence of 10^{12} phage ghosts per ml., not sedimentable on centrifugation at 30,000 g for 1 hr. after digestion with deoxyribonuclease, but was precipitated by 5 % trichloracetic acid. This protein impurity in the A preparation seems to be identical with one of the proteins that Hershey (1955) has shown to be released from the T2 bacteriophage head by osmotic shock. He has also shown that the non-sedimentable protein, which constitutes about 3 % of the total protein of the phage, may be injected with the deoxyribonucleic acid on infection but is not conserved in the progeny as is the nucleic acid. Our results suggest that part of this protein might play some role, perhaps an enzymatic one, in the initiation of phage protein synthesis. The confirmation of this and a study of the role of the deoxyribonucleic acid fractions in the synthesis of phage proteins awaits the isolation of the non-sedimentable proteins of the phage head in sufficient quantities.

DISCUSSION AND SUMMARY

If the hypothesis that the detailed structure of a deoxyribonucleic acid molecule can carry the information for the structure of a protein molecule is correct, the T2 bacteriophage-*Escherichia coli* system now presents the most suitable model system for testing it and investigating the relationship between the structures of the deoxyribonucleic acid and the related proteins, and the processes by which one is converted into the other. The fractionation work described above and elsewhere in this Symposium shows that deoxyribonucleic acid preparations from most sources are so complex that the isolation of a unique molecular species from them will be almost impossible. In the case of the T2 bacteriophage, a unique deoxyribonucleic acid species can be isolated which, present evidence suggests, carries genetic determinants controlling the structure of the tail protein responsible for the host-range properties. From the recent work of Kellenberger & Arber (1955), Williams & Fraser (1956) and Brown & Kozloff (1957) it seems likely that it will soon be possible to isolate this protein and work out the structural basis of host-range properties. Puck & Tolmach (1954) have shown that this probably depends on the spatial arrangement of free amino groups on part of the tail protein in the T2 bacteriophage.

The large size of the deoxyribonucleic acid molecule involved and the possibility that it carries determinants for several phage proteins makes a direct approach to the problem of the relation of nucleotide sequence and amino acid sequence very difficult, if not impossible. An indirect method, however, may be possible using the analysis of genetic fine structure as developed by Benzer (1955) and applied by Streisinger & Franklin (1956) to loci controlling host range in T2 phage. By using this method they have shown that the separation of the closest mutation sites on the genetic linkage group corresponds to the distance separating two to five nucleotides depending on the amount of T2 phage-deoxyribonucleic acid assumed to be carrying the genetic determinants. A study of the correspondence between genetic changes at a host-range region with structural changes in the tail protein as suggested by Streisinger & Franklin (1956) might be expected to yield information about the relation between nucleotide sequence and protein structure, except that the nucleotides would be anonymous. Another way round the practical difficulties of analysing the large deoxyribonucleic acid molecule in T2 phage would be to use a smaller phage.

The related problem of the way in which the genetic information carried by the deoxyribonucleic acid molecules is translated into protein structure, can now be more easily studied using the fractions of T2 bacteriophage-deoxyribonucleic acid and the protoplast ghost system. The preliminary results show that the presence of a protein component of the phage head may be required. The requirements for the ribonucleoside diphosphates indicates that polyribonucleotide synthesis may be necessary although this may be due to their action as co-factors in protein synthesis. In any case, the possibility that a specific structure, either ribonucleic acid or protein, is formed as an intermediate between the deoxyribonucleic acid and the final protein structure can now be readily tested.

REFERENCES

ANDERSON, T. F. (1953). *Ann. Inst. Pasteur*, **84**, 5.

AVERY, O. T., McLEOD, C. M. & McCARTY, M. (1944). *J. Exp. Med.* **79**, 137.

BENDICH, A. (1952). In *The Chem. and Physiol. of the Nucleus, Exp. Cell Res.*, Suppl. 2. New York: Academic Press.

BENDICH, A., FRESCO, J. R., ROSENKRANZ, H. S. & BEISER, S. M. (1955). *J. Amer. Chem. Soc.* **77**, 3671.

BENDICH, A., PAHL, H. B. & BEISER, S. M. (1956). *Cold Spr. Harb. Symp. Quant. Biol.* **21**, 31.

BENZER, S. (1955). *Proc. Nat. Acad. Sci., Wash.* **41**, 344.

BRACHET, J. (1941). *Arch. Biol.* **53**, 207.

BRACHET, J. (1955). In *The Nucleic Acids*, 2, 475. Ed. Chargaff, E. & Davidson, J. N. New York: Academic Press.

BRACHET, J. & CHANTRENNE, H. (1956). *Cold Spr. Harb. Symp. Quant. Biol.* **21**, 329.

BRENNER, S. (1957). (Private communication.)

BRENNER, S. & STENT, G. S. (1955). *Biochim. Biophys. Acta*, **17**, 473.

BROWN, D. D. & KOZLOFF, L. M. (1957). *J. Biol. Chem.* **225**, 1.

BROWN, G. L. & MARTIN, A. V. (1955). *Nature, Lond.* **176**, 971.

BROWN, G. L. & WATSON, M. (1953). *Nature, Lond.* **172**, 339.

CAMPBELL, D. H., LEUSCHER, E. & LERMAN, L. S. (1951). *Proc. Nat. Acad. Sci., Wash.* **37**, 575.

CASPERSSON, T. (1941). *Naturwissenschaften*, **29**, 33.

CAVALIERI, L. F. (1952). *J. Amer. Chem. Soc.* **74**, 1242.

CHARGAFF, E. (1955). In *The Nucleic Acids*, 1, 307. Ed. Chargaff, E. & Davidson, J. N. New York: Academic Press.

CHARGAFF, E., CRAMPTON, C. F. & LIPSHITZ, R. (1953). *Nature, Lond.* **172**, 289.

COLE, R. D., COOTE, J. & WORK, T. S. (1957). *Nature, Lond.* **179**, 199.

CRAMPTON, C. F., LIPSHITZ, R. & CHARGAFF, E. (1954a). *J. Biol. Chem.* **206**, 499.

CRAMPTON, C. F., LIPSHITZ, R. & CHARGAFF, E. (1954b). *J. Biol. Chem.* **211**, 125.

CRESTFIELD, A. M., SMITH, K. C. & ALLEN, F. W. (1955). *J. biol. Chem.* **216**, 185.

CRICK, F. H. C., GRIFFITH, J. S. & ORGEL, L. E. (1957). *Proc. Nat. Acad. Sci.,* *Wash.* **43**, 457.
CRICK, F. H. C. & WATSON, J. D. (1954). *Proc. Roy. Soc.* B, **223**, 80.
DE MARS, R. I. (1955). *Virology,* **1**, 83.
DE MOSS, J. A. & NOVELLI, G. D. (1956). *Biochim. Biophys. Acta,* **22**, 49.
FRANKLIN, R. E. & GOSLING, R. G. (1953). *Nature, Lond.,* **171**, 740.
GALE, E. F. & FOLKES, J. P. (1955a). *Biochem. J.* **59**, 661.
GALE, E. F. & FOLKES, J. P. (1955b). *Biochem. J.* **59**, 675.
GAMOW, G. (1954). *Nature, Lond.* **173**, 318.
GAMOW, G., RICH, A. & YČAS, M. (1955). *Adv. Biol. Med. Physics,* **4**. New York: Academic Press.
GAMOW, G. & YČAS, M. (1955). *Proc. Nat. Acad. Sci., Wash.* **41**, 1011.
GANDELMAN, B., ZAMENHOF, S. & CHARGAFF, E. (1952). *Biochim. Biophys. Acta,* **9**, 399.
GRUNBERG-MANAGO, M., ORTIZ, P. J. & OCHOA, S. (1956). *Biochim. Biophys. Acta,* **20**, 269.
HERSHEY, A. D. (1953). *J. Gen. Phys.* **37**, 1.
HERSHEY, A. D. (1955). *Virology,* **1**, 108.
HERSHEY, A. D. & CHASE, M. (1952). *J. Gen. Phys.* **36**, 39.
HERRIOTT, R. M. & BARLOW, J. L. (1952). *J. Gen. Phys.* **36**, 17.
HOAGLAND, M. B. (1955). *Biochim. Biophys. Acta,* **16**, 288.
HOAGLAND, M. B., KELLER, E. B. & ZAMECNIK, P. C. (1956). *J. Biol. Chem.* **218**, 345.
HOAGLAND, M. B., ZAMECNIK, P. C. & STEPHENSON, M. L. (1957). *Biochim. Biophys. Acta,* **24**, 215.
HOFFPAUIR, C. L. & GUTHRIE, J. D. (1950). *Text. Res. (J.)* **20**, 617.
HOTCHKISS, R. D. (1955). In *The Nucleic Acids,* **2**, 435. Ed. Chargaff, E. & Davidson, J. N. New York: Academic Press.
KAY, E. R. M., SIMMONS, N. S. & DOUNCE, A. L. (1952). *J. Amer. Chem. Soc.* **74**, 1724.
KELLENBERGER, F. & ARBER, W. (1955). *Z. Naturf.* **10b**, 698.
LANNI, F. & LANNI, Y. T. (1953). *Cold Spr. Harb. Symp. Quant. Biol.* **18**, 159.
LANNI, Y. T. (1954). *J. Bact.* **67**, 640.
LEDERBERG, J. (1956). *Proc. Nat. Acad. Sci., Wash.* **42**, 574.
LERMAN, L. S. (1955). *Biochim. Biophys. Acta,* **18**, 132.
LEVINTHAL, C. (1955). *R.C. Ist. lombardo,* **89**, 192.
LEVINTHAL, C. (1956). *Proc. Nat. Acad. Sci., Wash.* **42**, 394.
LEVINTHAL, C. & THOMAS, C. A. (1957a). *Biochim. Biophys. Acta,* **23**, 453.
LEVINTHAL, C. & THOMAS, C. A. (1957b). In *The Chemical Basis of Heredity.* Ed. McElroy, W. D. & Glass, B. Baltimore: Johns Hopkins Press.
LIPSHITZ, R. & CHARGAFF, E. (1956). *Biochim. Biophys. Acta,* **19**, 256.
LUCY, J. A. & BUTLER, J. A. V. (1954). *Nature, Lond.* **174**, 32.
MAHLER, H. R. & FRASER, D. (1956). *Biochim. Biophys. Acta,* **22**, 197.
MCQUILLEN, K. (1956). In *Bacterial Anatomy. Soc. Gen. Microbiol. 6th Symp.* Ed. Spooner, E. T. C. & Stocker, B. A. D. Cambridge Univ. Press.
MIRSKY, A. E., OSAWA, S. & ALLFREY, V. G. (1956). *Cold Spr. Harb. Symp. Quant. Biol.* **21**, 49.
PETERSON, E. A. & SOBER, H. A. (1956). *J. Amer. Chem. Soc.* **78**, 751.
PUCK, T. & SAGIK, B. (1953). *J. Exp. Med.* **97**, 807.
PUCK, R. & TOLMACH, L. J. (1954). *Arch. Biochem. Biophys.* **51**, 229.
RAPSKE, R. (1956). *Biochim. Biophys. Acta,* **22**, 189.
SALTON, M. R. J. & MCQUILLEN, K. (1955). *Biochim. Biophys. Acta,* **17**, 465.
SHOOTER, K. V. & BUTLER, J. A. V. (1955). *Nature, Lond.* **175**, 500.

SINSHEIMER, R. L. (1954). *Science*, **120**, 551.
SINSHEIMER, R. L. (1956). *Proc. Nat. Acad. Sci., Wash.* **42**, 502.
SPIEGELMAN, S. (1957). In *The Chemical Basis of Heredity*. Ed. McElroy, W. D. & Glass, B. Baltimore: Johns Hopkins Press.
STREISINGER, G. (1956). *Virology*, **2**, 388.
STREISINGER, G. & FRANKLIN, N. C. (1956). *Cold Spr. Harb. Symp. Quant. Biol.* **21**, 103.
STREISINGER, G. & WEIGLE, J. (1956). *Proc. Nat. Acad. Sci., Wash.* **42**, 504.
TOMCSIK, J. & GUEX-HOLZER, S. (1952). *Schweiz. Z. Allg. Path.* **15**, 517.
VOLKIN, E. & ASTRACHAN, L. (1956). *Virology*, **2**, 149.
WEIBULL, C. (1953a). *J. Bact.* **66**, 688.
WEIBULL, C. (1953b). *J. Bact.* **66**, 696.
WEIBULL, C. (1956). In *Bacterial Anatomy. Soc. Gen. Microbiol. 6th Symp.* Ed. Spooner, E. T. C. & Stocker, B. A. D. Cambridge Univ. Press.
WEIDEL, W. (1951). *Z. Naturf.* **6b**, 251.
WILKINS, M. H. F. (1956). *Cold Spr. Harb. Symp. Quant. Biol.* **21**, 75.
WILKINS, M. H. F., STOKES, A. R. & WILSON, H. R. (1953). *Nature, Lond.* **171**, 738.
WILLIAMS, R. C. & FRASER, D. (1956). *Virology*, **2**, 388.
WILSON, H. R. & WILKINS, M. H. F. (1955). (Private communication.)
ZINDER, N. D. & ARNDT, W. F. (1956). *Proc. Nat. Acad. Sci., Wash.* **42**, 586.

STUDIES OF DEOXYRIBONUCLEIC ACIDS WITH THE AID OF ANION EXCHANGERS

By AARON BENDICH, HERBERT B. PAHL,* HERBERT S. ROSENKRANZ† AND MORTON ROSOFF

The Sloan-Kettering Division of Cornell University
Medical College, New York 21, New York

It is probably no accident that the deoxyribonucleic acids (DNA) which have been isolated by different methods from a wide variety of different sources have been found to consist of mixtures of polynucleotides. The metabolic, chemical and physico-chemical heterogeneity of DNA from many sources (including viruses) has now been demonstrated by several independent techniques (Bendich, 1952; Bendich, Russell & Brown, 1953; Sherratt & Thomas, 1953; Chargaff, Crampton & Lipshitz, 1953; Brown & Watson, 1953; Lucy & Butler, 1954; Hotchkiss & Marmur, 1954; Chargaff, 1955; Bendich, Fresco, Rosenkranz & Beiser, 1955; Ephrussi-Taylor, 1955; Brown & Martin, 1955; Bendich, Pahl & Beiser, 1956; Shooter & Butler, 1955, 1956; Zamenhof, Reiner, De Giovanni & Rich, 1956; Levinthal, 1956; Kawade & Watanabe, 1956). The diversity of the macro-molecular polynucleotide components of DNA seems to be a reflexion of the diversity of the genetic information carried by the cell. There is abundant evidence which supports the concept that the genetic determinants of the cell are composed of DNA (Hotchkiss, 1955; Zamenhof, 1956a, b; Sinsheimer, 1957). From studies of the fractionation of pneumococcal transforming DNA on columns of the anion exchanger ECTEOLA, evidence has been obtained for the heterogeneity of this biologically active material (Bendich et al. 1956; Pahl, Beiser & Bendich, 1957). It is the purpose of this paper to discuss a number of problems which bear on the question of the biological replication of these macromolecules and other applications to which the chromatographic fractionation procedure using ECTEOLA has been put.

CHROMATOGRAPHIC METHOD

The fractionation technique‡ employed in our studies involves the use of the anion exchanger ECTEOLA which is prepared from cellulose,

* Fellow of the National Cancer Institute of the Public Health Service 1955 to 1957. Present address, Department of Biochemistry, Vanderbilt University, Nashville, Tenn.
† Alfred P. Sloan Foundation Pre-doctoral Fellow.
‡ A paper describing the details of this technique is in preparation.

epichlorohydrin and triethanolamine as described by Peterson & Sober (1956). Specimens of DNA (2–3 mg.) were ordinarily dissolved either in 0·01 M neutral phosphate buffer or in 0·001 M-NaCl solution and applied to columns (about 1·0 × 5·0 cm.) of ECTEOLA. On elution with solutions of NaCl of gradually increasing molarity followed by NaCl solutions of gradually increasing pH, DNA is quantitatively (95–104 %) recovered as numerous chromatographic peaks (Bendich *et al.* 1955; Bendich *et al.* 1956; cf. Astrachan & Volkin, 1957). A typical chromatographic profile is presented in Fig. 1, which shows the results obtained with a sample of

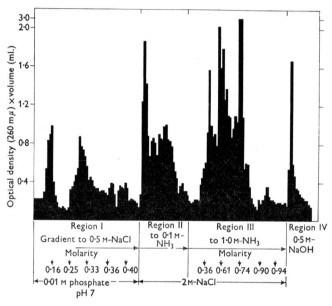

Fig. 1. Chromatography on a column of ECTEOLA (0·8 × 5·2 cm.; 0·20 m-equiv./g.; flow-rate, about 4 ml./hr.) of 3 mg. of DNA isolated from human leucocytes by a detergent method (Kay, Simmons & Dounce, 1952). Collections were made every 2 hr. Recovery, 96 %. The black area is directly proportional to the quantity of DNA. The leucocytes were obtained from a patient with acute myeloblastic leukaemia (diagnosis kindly provided by Dr L. D. Hamilton). We are indebted to Dr E. E. Polli for this sample of DNA.

DNA of human white cells (diagnosis, acute myeloblastic leukaemia). The amount of DNA present in each column collection was determined by measurement of the ultra-violet absorption (at 260 mμ). Four distinct elution regions, as shown on the abscissa, were employed in these experiments: I (a continuous gradient increase from 0·0 to 0·5 M-NaCl); II (0·1 M-NH$_3$ in 2·0 M-NaCl); III (a gradient increase from 0·1 M to 1·0 M-NH$_3$, in 2·0 M-NaCl) and IV (0·5 M-NaOH). For convenience, these regions will be referred to by number in the sections which follow.

Reproducibility. The chromatographic procedure has been found to be highly reproducible. Examples of the reproducibility of the chromatographic patterns obtained with a number of different specimens of DNA have been presented previously (Bendich *et al.* 1956). Still another example is shown in Fig. 2 in which it is seen that much of the fine structure in the profiles can be reproduced in great detail (compare the 2- and 4-week patterns, for example). The high degree of reproducibility achieved with

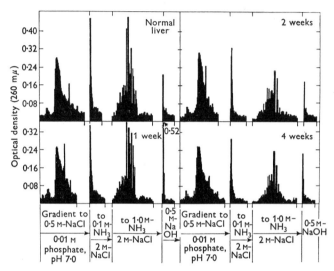

Fig. 2. Effect of partial hepatectomy on liver DNA. Chromatography of normal and regenerating liver DNA, 1, 2 and 4 weeks after the operation. The four columns each contained 0·5 g. of ECTEOLA (0·20 m-equiv./g.) and 3 mg. of a DNA and were run simultaneously. The flow-rate was carefully maintained at 4 ml./hr., with collections made at 2-hr. intervals. Recovery, 96–103 %.

this method provides much support for the conclusion that the differences noted (Bendich *et al.* 1956) in the chromatographic patterns of DNA either from different sources or from such closely related sources as T6r and T6r+ bacteriophage, or the brain and kidney of the rat are a reflexion of intrinsic differences within the specimens of the DNA. At the same time, it must be kept in mind that the problem of the 'nativeness' or the degree to which the isolated DNA is biologically intact is a serious one, and one with which investigators must always be concerned whenever macromolecular polyelectrolytes are isolated from living tissue. Most investigators today are in agreement that, regardless of the source of the DNA or the method of isolation employed, DNA is heterogeneous. It is difficult to determine how much the procedure of isolation contributes to the observed heterogeneity, and it may be that the application of the chromatographic method to this question will materially aid in its clarification.

Resolution. The ECTEOLA-chromatography procedure appears to possess a reasonably high degree of resolution as shown by the following observations: (1) The method has been found capable of resolving DNA molecules which contain the thymine-analogue 5-bromouracil from those which lack this base (Bendich, Pahl & Brown, 1957). (2) Column fractions of calf-thymus DNA obtained with this procedure were found to contain different relative amounts of the heterocyclic bases. Perhaps the most striking illustration of the differences in base composition was obtained when the ratios adenine/thymine and guanine/cytosine for many of the column fractions were found to be significantly different from the frequently obtained value of 1·0 (Bendich *et al.* 1956). (3) Other indications of the degree of resolution inherent in this procedure were obtained following the chromatographic fractionation of pneumococcal transforming DNA. Fractions were obtained which showed a variation of more than fiftyfold in the relative specific activity for transformation to streptomycin resistance (Bendich *et al.* 1956). Furthermore, a partial separation of the strepto-mycin, penicillin and mannitol activities has been reported (Pahl *et al.* 1957).

A technical question closely allied to the problem of column resolution is that of rechromatography. Evidence has been obtained that when a selected narrow portion of a chromatographic peak is rechromatographed, the large bulk of the DNA appears at that place in the chromatogram (in terms of the particular ionic strength and pH) at which the original specimen was taken (Bendich *et al.* 1956).

PHYSICO-CHEMICAL BASIS OF THE SEPARATION

It is pertinent to determine the basis by which the separations of the poly-nucleotide components are achieved when DNA is chromatographed on ECTEOLA. Two of the considerations which have been investigated to date are: (1) the base compositions of the various column fractions, and (2) the determination of the shape of the chromatographic profile as a function of the molecular size of the DNA sample. These points are discussed below.

Base composition. One likely basis for the fractionation is the difference in base composition which is observed among the various column fractions (Bendich *et al.* 1956). In this connexion, it should be noted that other DNA fractionation procedures which are presumably based upon other principles have also yielded fractions with different base compositions (Brown & Watson, 1953; Lucy & Butler, 1954; Chargaff, 1955; Bendich, 1956).

Molecular size. In an experimental approach to this question, studies were carried out using both oligonucleotide mixtures and heat-treated DNA, the results in each case being compared with the chromatographic pattern obtained with a standard preparation of highly polymerized calf-thymus DNA.

(1) *Oligonucleotide experiments.* It was found that nucleotides of low molecular weight were either poorly retained by columns of ECTEOLA or were eluted by neutral salt solutions of low ionic strength. For example, mono-deoxyribonucleotides were quantitatively eluted by means of neutral 0·01 M phosphate buffer. However, solutions of about 0·17–0·19 M-NaCl (in the phosphate buffer) were required for the quantitative elution of the oligonucleotides present in deoxyribonuclease digests of calf-thymus DNA (Bendich *et al.* 1956). Since the largest nucleotide in such a digest is of the order of an octanucleotide (i.e. with a molecular weight of about 2400; Sinsheimer, 1954), it was concluded that fractions eluted with solutions of higher ionic strength or increased pH were larger. A number of physico-chemical studies were therefore carried out to determine the extent to which the chromatographic selection was due to differences in molecular size, state of aggregation or shape.

(2) *Sedimentation studies.* It had been found previously (Bradley & Rich, 1956) that the fractions of ribonucleic acid obtained with columns of ECTEOLA showed increases in sedimentation constant (hence, presumably, molecular weight) which could be correlated with the increases in the ionic strength of the eluents. Ion-exchange chromatography of yeast-ribonucleic acid on columns of Dowex-2 has yielded fractions which show a positive correlation between their molecular weight and the ionic strength of the eluting solution (Miura & Suzuki, 1956). Those ribonucleic acids examined, however, possess molecular weights which are considerably smaller than those of the DNA specimens ordinarily studied, and also exhibit lower sedimentation coefficients than DNA (Bradley & Rich, 1956; Shooter & Butler, 1955, 1956). It was of interest, therefore, to determine whether there was also a correlation between the sedimentation coefficient or molecular weight of DNA and the ionic strength or pH of the eluting fluid.

Ultra-centrifugal analyses of calf-thymus DNA have shown a wide distribution in sedimentation coefficient (Shooter & Butler, 1956; Schumaker & Schachman, 1957). A specimen of the standard unfractionated calf-thymus DNA used in the present studies also showed a similar distribution (Fig. 3). Assuming a free draining coil, these sedimentation data can be interpreted as corresponding to a range of molecular weights of about 1×10^6 to 25×10^6. This is consistent with the weight-average molecular weight of $5·5 \times 10^6$ found for this sample by light scattering (Cavalieri, Rosoff & Rosenberg, 1956). In contrast, chromatographic fractions of the standard DNA from region I (fractions 1 and 2, Fig. 3) exhibit a much narrower distribution of sedimentation coefficient. These fractions (Fig. 3) also show much smaller weight-average sedimentation coefficients (11·2 and 12·8, respectively) than the original unfractionated DNA (17·8). These

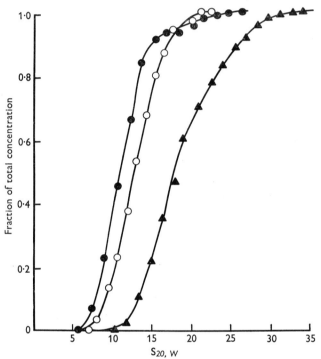

Fig. 3. Integral distribution of sedimentation coefficients. The unfractionated calf-thymus DNA and the column fractions were equilibrated by dialysis against 0·2M-NaCl. Solutions of approximately 0·003 % were examined in a Spinco Model E ultra-centrifuge equipped with ultra-violet absorption optics (Schumaker & Schachman, 1957; Shooter & Butler, 1956). ●-●-●, Fraction 1 (6% of the original DNA) was eluted in region I by 0·16–0·24M-NaCl; O-O-O, fraction 2 (10% of the original DNA) was eluted in region I by 0·25–0·34M-NaCl; ▲-▲-▲, unfractionated calf-thymus DNA.

results parallel those found for ribonucleic acid fractions (Bradley & Rich, 1956). Fractions 1 and 2 together comprise only about 16% of the original DNA, and it is therefore to be expected that fractions in the remaining chromatographic regions will show higher sedimentation coefficients.*

(3) *Heating experiments.* The effect of heat on solutions of DNA has been studied by many investigators in attempts to discern the character of the macromolecule when in solution. Accordingly, specimens of calf-

---

* *Footnote added in proof:* This expectation has been realized in experiments carried out since this manuscript was submitted.

thymus DNA (prepared according to Schwander & Signer, 1950) were heated for one hour at 100° C. (1 mg./ml.), and subjected to chromatographic analysis on columns of ECTEOLA. The chromatographic profiles are shown in Fig. 4 together with that of an unheated specimen. The chromatograms are clearly different from that of the untreated DNA since the bulk of the DNA heated in water was recovered in region I, whereas the heating in 0·01 M neutral phosphate solution gave rise to a product which on chromatography showed hardly any of the first large peak characteristic of the unheated DNA. Since the amount of DNA in each column collection is directly proportional to the black area shown in this plot (Fig. 4), it can be seen that the heating of the DNA in the 0·01 M salt solution resulted in almost a quantitative loss of the first peak and in a considerable loss of the DNA of elution region III (i.e. in the 0·1–1·0 M-NH$_3$/2 M-NaCl range). These fractions were obtained in region II, presumably together with that material ordinarily eluted in this region.

These differences provide evidence for an important aspect of the physicochemical basis of the chromatographic separation because they can be interpreted in terms of the data which have been reported by other investigators for DNA treated in a similar fashion. The heating of DNA in water or in solutions of very low ionic strength leads to a considerable decrease in sedimentation coefficient, viscosity and molecular weight (Dekker & Schachman, 1954; Schachman, 1957a, b; Doty, 1955; Doty & Rice, 1955; Sadron, 1955; Butler & Shooter, 1957; Pouyet & Weill, 1957). When heated in salt-free water solution, the sedimentation coefficient, for example, of T2 phage DNA, drops from a value of 20–40 to about 4–6·5 (Butler & Shooter, 1957).

The molecular weight and sedimentation behaviour of Schwander-Signer calf-thymus DNA have been studied by many investigators. Since the chromatogram of the unheated DNA (Fig. 4, upper profile) shows the largest amount of DNA to be in region III, it can be expected that this region contains much of that larger molecular-weight material which is almost quantitatively lost upon heating in water, and which subsequently is recovered in the first peak (region I). Since DNA heated in water contains polynucleotide material of rather small size, it can be inferred that fractions eluted in region I are also small. The position of a particular DNA peak in a chromatographic region may therefore be a sensitive indicator of molecular size.

The interpretation of the results of the heating of DNA in the salt solution is complicated by apparently conflicting results in the literature. Some investigators have reported a decrease in sedimentation coefficient after such treatment, whereas others have noticed an increase (presumably due to aggregation). The chromatographic pattern (Fig. 4, bottom profile) would seem to be consistent with both interpretations, since DNA is lost to region II from both regions I and III.

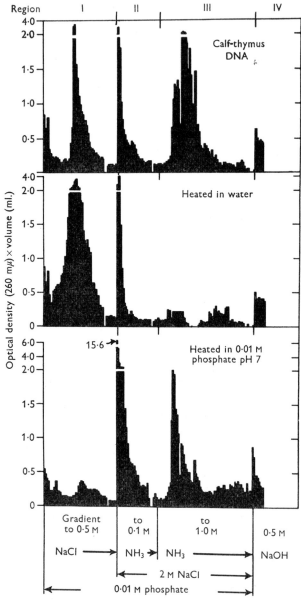

Fig. 4. Chromatography of calf-thymus DNA after heating in distilled water (centre profile) and o·o1 M phosphate buffer, pH 7 (bottom profile). Upper profile is a chromatogram of the unheated calf-thymus DNA. Each column (o·8 × c. 5·3 cm.) contained o·5 g. of ECTEOLA (o·20 m-equiv./g.). Recovery in each case, 100 ± 5 %.

If fractions of DNA obtained in region I in an ECTEOLA chromatogram are smaller than the average, whereas those in region III are much larger (some evidence for these suppositions has been presented above), it could then be said, as one possibility, that the heating in salt solution had apparently resulted in an aggregation of most of the smaller DNA particles and a degradation of some of the larger polynucleotides present in the original DNA. Doty & Rice (1955), have observed an increase of approximately 50 % in the sedimentation coefficient of calf-thymus DNA after heating in 0·2 M-NaCl solution at 100° C. On the other hand, Shooter, Pain & Butler (1956), and Butler & Shooter (1957), have found a limited decrease in sedimentation constant together with a great decrease in viscosity. They interpreted these results as possibly due to a decrease of molecular weight (to $\frac{1}{4}$ or $\frac{1}{6}$ of the original). The effects of the heating of DNA in salt solutions appear to depend upon the concentration of the DNA (Shooter et al. 1956) as well as on the ionic strength of the salt solution (Sadron, 1955; Pouyet & Weill, 1957). It must be borne in mind, however, that a variety of different approaches, including the ion-exchange chromatographic technique, have clearly shown that preparations of DNA, as studied, are mixtures of different polynucleotides. It is therefore pertinent to ask whether conclusions drawn from physico-chemical experiments on unfractionated DNA should not be tempered by this fact.* It may be that all of the component polynucleotides in a preparation of DNA are not affected equally by treatment with heat. The results of the heating of DNA in 0·01 M-phosphate in the present study are at least consistent with this interpretation. However, it is clear that a *less* drastic alteration accompanies the heating of DNA in salt than in distilled water solution. This supports the conclusion of Sadron (1955) that salt exerts a limited protective effect against the thermal breakdown of DNA. The presence of salt also protects against the loss of biological activity which otherwise occurs when transforming DNA is stored in solutions of very low ionic strength either at ordinary or ice-box temperatures (Avery, MacLeod & McCarty, 1944; Zamenhof, Alexander & Leidy, 1953; Thomas, 1954; Cavalieri, Rosoff & Rosenberg, 1956; Doty, 1957 a, b).

It may not be without significance that the peak in region II (0·1 M-NH$_3$/2 M-NaCl) of the chromatogram of the water-heated DNA (centre profile, Fig. 4) appears to be very similar in size and shape to the corresponding one for the unheated specimen. Experiments are under way to determine whether these two peaks are actually composed of DNA

---

* It would be unwise to infer that DNA fractions obtained by any current fractionation procedure are themselves homogeneous. This points to the need for substantial improvement in existing methods.

fractions which possess similar properties. One exciting possibility is that there may exist in DNA a fraction that is resistant to heating in distilled water (or, perhaps, to other forms of mistreatment such as subjection to X- or ultra-violet irradiation, nitrogen mustard and other mutagens, etc.).

APPLICATIONS TO BIOLOGICAL PROBLEMS

Regenerating rat liver. In an experiment designed to study the effect of partial hepatectomy on the chromatographic profile of DNA from regenerating rat liver as compared with that from the normal liver, male Wistar rats were partially hepatectomized and the DNA isolated after one, two and four weeks. The DNA was prepared from the nuclear fraction (Hogeboom, Schneider & Striebich, 1952) by repeated extraction with cold 10 % sodium chloride-0·05 M-citrate.

The results of these studies are shown by the four chromatographic profiles reproduced in Fig. 2. Because two recognizable fractions can be isolated from liver shortly after partial hepatectomy by extraction with hot saline (Bendich *et al.* 1953), one of which disappears following regeneration (compare Diermeier, Di Stephano & Bass, 1955), it might be anticipated that differences between the chromatographic profiles of the normal and regenerating liver-DNA would be observed. That this was obviously not so may indicate that those particular differences which exist between the two types of hot saline-extracted DNA in regenerating liver are probably not the same as those which serve as the basis for the chromatographic separation under the particular conditions employed in the present studies.

In these experiments, quantitative measurements were made of the amount of DNA actually isolated as a percentage both of the DNA originally present in the tissue and in the isolated nuclear fraction. About 83 % of the tissue DNA was recovered in the isolated nuclear fraction. The isolated DNA represented 47 % of the DNA of this nuclear fraction (or 39 % of the fresh-tissue DNA). Thus the DNA actually isolated was only a portion of the total DNA in the tissue, and to this extent the chromatographic profiles reflect only this part of the DNA. However, in view of the close similarity of the profiles of the four DNA samples (Fig. 2), it should be realized that either the extraction procedure was reproducible despite the fact that only a portion of the DNA was isolated, or else the isolated specimen did indeed represent a true sampling of the entire tissue DNA.

Escherichia coli-DNA containing 5-bromouracil. The incorporation of 5-bromouracil into the DNA of *E. coli* by equimolar replacement of the thymine of the nucleic acid has been the subject of a number of studies (Dunn & Smith, 1954; Zamenhof *et al.* 1956*a*, *b*). The chromatographic fractionation of *E. coli*-DNA in which about 50% of the thymine had been replaced by 5-bromouracil has been reported (Bendich *et al.* 1957). The results of the latter study indicated that this thymine analogue was unequally incorporated into the various DNA molecules. The biological significance of this finding has been discussed (*loc. cit.*). Thus a discrimination between DNA which contains 5-bromouracil and DNA which does not has been achieved by ECTEOLA-chromatography. This indicates the possibility that other differences among DNA molecules can also be detected by this procedure, and hence this makes available an experimental approach to problems of a similar nature. For example, the metabolic heterogeneity of chicken bone-marrow-DNA with respect to the incorporation of thymidine-^{14}C has been demonstrated by Friedkin & Wood (1956) with columns of ECTEOLA.

5-Bromouracil is only one example of purine and pyrimidine analogues which either have been or probably will be found to be incorporated into the DNA structure. It may thus be anticipated that fractionation experiments similar to those mentioned in this paper will prove of value in attempts to understand more fully the biological significance of the metabolic and structural differences which undoubtedly exist among the deoxyribonucleic acids of various sources.

Pneumococcal transforming DNA. The observation that different biological activities can be separated to some degree by the fractionation of pneumococcal transforming DNA on columns of ECTEOLA has been the subject of recent reports (Bendich *et al.* 1956; Pahl *et al.* 1957). Some of the more pertinent findings which have been obtained from such experiments include the following: (*a*) there appears to be no significant destruction of the biologically active DNA as a result of its adsorption to and elution from ECTEOLA; (*b*) some of the column fractions obtained show more, while others show less specific biological activity than does the starting, unfractionated DNA; (*c*) when the ratio of two or more activities is followed in individual column collections, it is found that such ratios show significant differences—this indicates, presumably, that a fractionation of different biological characters is involved; (*d*) in the case of the transformation to streptomycin resistance, the sum of the activities of the various fractions was about 70% greater than the activity applied to the column (see below for interpretation); (*e*) assays both of unfractionated DNA and of individual column collections indicate that the *quality* of each is about the same—in the sense in which it is used here, the word 'quality' means the relative extent of degradation of the sample in question (see below).

With regard to point (*d*), the recovery of streptomycin transforming

DNA from the column, the data indicate that when the *unfractionated* DNA is assayed at non-saturation levels the full biological activity for some reason cannot be expressed. One interpretation is that when the unfractionated DNA is used, there is a competitive inhibition by those DNA molecules which do not possess the activity, whereas with the column fractions the competition occurs to a lesser extent, presumably as a result of an alteration in the ratio of the active to the non-active DNA.

INTEGRITY OF THE DNA FRACTIONS

It is very important to determine whether the chromatography on ECTEOLA alters the integrity of the DNA. Three independent lines of evidence are cited below which support the contention that no extensive alteration occurs in DNA when fractionated on ECTEOLA.

The first line of evidence came from studies dealing with the rechromatography of column fractions (see above). In those experiments (Bendich *et al.* 1956), it was found that the bulk of a given fraction when subjected to rechromatography appeared at the same position in the elution profile as had the original. Consequently, if alteration of the DNA had attended its initial fractionation on ECTEOLA, this did not materially influence its chromatographic properties. In this regard, the chromatographic profiles obtained with heated DNA (Fig. 4) are of significance since there is clearcut evidence from these experiments for a change in the shape of the profile when DNA is deliberately subjected to a treatment known to denature or alter the nucleic acid.

A second line of evidence is that, following the fractionation of pneumococcal transforming DNA, essentially all of the biological activity may be recovered (Pahl *et al.* 1957). These data take on added significance in view of the many studies (see Zamenhof, 1957) which indicate that the loss of biological activity is perhaps the most sensitive criterion for the occurrence of physico-chemical change ('denaturation') in DNA molecules.

The third set of data is concerned with the activity-response curves which were obtained with both unfractionated and fractionated transforming DNA. Hotchkiss (1957) has pointed out recently that '...the saturation yield of transformants is a measure of the quality of a DNA preparation...'. Accordingly, the total number of transformed cells obtained with *saturation levels* of DNA was studied, with the result that no significant differences for either the unfractionated sample or for the subfractions recovered from the ECTEOLA column were observed (Pahl *et al.* 1957). It thus would appear that the quality of the DNA in an individual column collection is, by this criterion, the same as that of the unfractionated

DNA. These data therefore lead to the conclusion that the column fractions are probably no more degraded or denatured than the unfractionated sample.

GENERAL REMARKS

Replication of DNA. Since DNA, as ordinarily isolated, consists of a mixture of molecules of varying size and base composition, they most likely differ among themselves in structural arrangement. This is consistent with the notion that the large number of different biological characters in a given species are presumably determined by molecules of DNA of different chemical and/or physical properties. This does not mean, of course, that each genetic activity is associated with just one DNA molecule, nor does it mean necessarily that there must be as many different DNA molecules as there are genes. Because DNA is a huge macromolecule, it is easy to visualize how each molecule could show many genetic properties. An interesting calculation made by Zamenhof (1957) has indicated that about 200 molecules of DNA are present in a single *Haemophilus influenzae* cell. As Zamenhof pointed out, this number of DNA molecules would appear to be too small to determine all of the hereditary characters of the cell unless it is postulated either (*a*) that not all the characters are determined by DNA, or (*b*) that one molecule of DNA determines several characters. The extent to which doubly marked pneumococcal transformants occur when appropriate DNA preparations are employed has been regarded as evidence for the existence of linked genetic factors in pneumococcal DNA (Hotchkiss & Marmur, 1954).

Similarly, the observation that DNA-mediated multiple capsule transformation can occur in a strain of *H. influenzae* was made by Leidy, Hahn & Alexander (1953). Furthermore, the results of the chromatographic fractionation of the transforming principle of *Pneumococcus* (Bendich *et al.* 1956; Pahl *et al.* 1957) indicate that many different fractions exhibit qualitatively the same type of activity, although they vary quantitatively in transforming activity. A picture which emerges from these various studies is that the total DNA of a single source such as the *Pneumococcus* may consist of a large number of differently constituted molecules, many of which carry several genetic activities. Another view is that a particular type of genetic activity may be associated with different DNA molecules.

In this context, it is of interest to consider whether a complex mixture of such complicated material can be replicated exactly by the cell after each division. Since biological identity is almost always maintained, the genetic material must, to this extent, also retain its identity in succeeding generations of cells. For example, if genetic continuity requires exact replication of DNA, one might consider that 'exact' replication of the 2- and 4-week post-hepatectomy liver DNA had actually occurred (see the chromatographic profiles in Fig. 2). Even though these diagrams are very nearly the same, it can be seen that in fact they are not so in every detail. This raises the tantalizing question, how different can two things be and still be the same, or how similar can two things be and still be different? A corollary question is how much change is possible before a significant genetic alteration is detectable? It is felt that progress in these problems requires the use of fractionated DNA.

Taxonomic considerations. In considering, from a purely taxonomic view-point, the various classifications which have been proposed for listing organisms according to their degree of specialization or evolutionary development, a difficulty is encountered when an attempt is made to include the viruses. This difficulty probably arises in large part from a lack of agreement on the terms 'living' and 'non-living'. (For interesting discussions concerning these terms see Pirie, 1938, 1952, 1954; Cohen, 1956.) In view of the fact that the *free* nucleic acids are active both in bacterial transformation (DNA; Hotchkiss, 1955), in producing viral progeny when applied to the leaves of the tobacco plant (RNA; Gierer & Schramm, 1956a, b; Fraenkel-Conrat, Singer & Williams, 1957) and in producing Eastern equine encephalomyelitis in mice (Wecker & Shäfer, 1957) it is proposed that these substances should be accorded taxonomic recognition along with the viruses and the fully competent cells. Viewed in this manner, a least common denominator for all members of the extended evolutionary scale is nucleic acid.

Although less is now known for RNA, a widely recurring if not charac-teristic feature of DNA is its heterogeneous nature. As noted previously (Bendich *et al.* 1956; Pahl *et al.* 1957), the DNA of *Pneumococcus* is partially separable by anion-exchange chromatography into polynucleotide fractions with different biological activities. In the event that these or other preparations of DNA can be resolved further into discrete molecules endowed with specific genetic activity, a further evolutionary subclassifi-cation might be possible. It is conceivable that this might help in the discernment of the chemical basis of heredity and it is hoped that refined anion-exchange chromatographic techniques might contribute towards this end.

Possible applications of the fractionation procedure. Many biological problems may be studied with the technique at its present level of develop-ment. Although it may be asking too much to hope for a method whereby truly homogeneous DNA fractions can soon be obtained (witness the problem of the microheterogeneity of proteins; Colvin, Smith & Cook, 1954), this is no reason why a bold approach should not be made upon such studies using the tools at hand. In the paragraphs below are outlined some of the many questions which might profitably be explored by the present chromatographic technique.

For example, one may expect that such an analytical tool might be useful in studying whether changes occur in the DNA of a bacterial population when the environmental conditions are altered, as in the thymine-deficiency experiments with *Escherichia coli* 15T- (Cohen, 1957). A similar problem is the chromatographic analysis of *E. coli*-DNA, the thymine of which is replaced by varying amounts of 6-methylaminopurine under various conditions of growth (Dunn & Smith, 1955). Since, in this instance, the thymine has been replaced not by a pyrimidine but rather by a purine, would such DNA be distinguishable from that containing 5-bromouracil? Would the DNA fractions vary in their content of 6-methylaminopurine? What are the properties and enzymic susceptibilities of the fractions which contain this base as compared with those which may lack it?

Another application of this procedure in which we have been interested is the incorporation of 5-bromouracil into the biologically active DNA of *Pneumococcus*, in order to study the correlation of the biological activity of a fraction with the 5-bromouracil content. In other words, to what extent can the thymine (or any other normally occurring constituent) be replaced without resulting in the loss or alteration of biological activity? Unfortunately, it has not yet been possible to incorporate this analogue into pneumococcal DNA, and so this particular problem still remains unsolved.

Since encouraging results have been obtained to date in the application of this chromatographic method to biologically active transforming DNA from *Pneumococcus*, it would appear that the procedure holds much promise for future investigations in this area of biochemical genetics. For example, it would be very interesting to subject the individual column fractions to nitrous acid, ionizing and ultra-violet irradiation, heat, extremes of pH, low ionic strength, and deoxyribonuclease, all of which are known to result in the loss of biological activity. Another interesting line of investigation is that which deals with the determination of the molecular size of the smallest biologically active DNA fraction. Experiments in this direction have already been initiated, although no definite conclusion can be drawn at this time. Does the molecular weight of the smallest biologically active DNA particle depend upon the particular activity being followed? Can genetically linked markers be separated? In the case of the transformation to drug resistance, it has been found that the resistance is acquired in discrete steps (Hotchkiss, 1955). Can the fractionation of a DNA which can confer a high level of resistance yield fractions which will transform non-resistant cells to varying levels of resistance?

The DNA from the white cells of patients with chronic lymphatic leukaemia has been found to be chromatographically different from that of the leucocytes from patients with chronic myeloid leukaemia (unpublished results, with E. E. Polli). Differences in the sedimentation constants of these materials have also been found (Shooter, K. V. & Polli, E. E., unpublished observations). It is not yet known whether such differences have a significant relationship to these diseases. The chromatographic procedure may afford a sensitive indication of the effect of certain forms of treatment (X-irradiation, alkylating agents, etc.) and might even have diagnostic and prognostic value.

This investigation was supported by funds from the American Cancer Society, National Cancer Institute, National Institutes of Health, Public Health Service (Grant CY-3190), and from the Atomic Energy Commission (Contract AT (30-1), 910).

The authors take pleasure in acknowledging the advice and valuable help of Dr George B. Brown, Dr Marion Barclay, Dr Sam M. Beiser, Dr Giampiero di Mayorca and Dr Elio E. Polli, and are highly indebted for the valuable assistance of Mrs Grace Korngold. It is also a pleasure to thank Dr Howard K. Schachman and Miss Sue Hanlon for advice concerning the ultra-centrifugal analyses.

REFERENCES

ASTRACHAN, L. & VOLKIN, E. (1957). *J. Amer. Chem. Soc.* **79**, 130.
AVERY, O. T., MacLEOD, C. M. & McCARTY, M. (1944). *J. Exp. Med.* **79**, 137.
BENDICH, A. (1952). *Exp. Cell. Res.* **3**, suppl. **2**, 181.
BENDICH, A. (1956). In *Essays in Biochemistry*, p. 14. Ed. Graff, S. New York: John Wiley.
BENDICH, A., FRESCO, J. R., ROSENKRANZ, H. S. & BEISER, S. M. (1955). *J. Amer. Chem. Soc.* **77**, 3671.
BENDICH, A., PAHL, H. B. & BEISER, S. M. (1956). *Cold Spr. Harb. Symp. Quant. Biol.* **21**, 31.
BENDICH, A., PAHL, H. B. & BROWN, G. B. (1957). In *The Chemical Basis of Heredity*, p. 378. Ed. McElroy, W. D. & Glass, B. Baltimore: Johns Hopkins Press.
BENDICH, A., RUSSELL, P. J., Jr. & BROWN, G. B. (1953). *J. Biol. Chem.* **203**, 305.
BRADLEY, D. F. & RICH, A. (1956). *J. Amer. Chem. Soc.* **78**, 5898.
BROWN, G. L. & MARTIN, A. V. (1955). *Nature, Lond.* **176**, 971.
BROWN, G. L. & WATSON, M. (1953). *Nature, Lond.* **172**, 339.
BUTLER, J. A. V. & SHOOTER, K. V. (1957). In *The Chemical Basis of Heredity*, p. 540. Ed. McElroy, W. D. & Glass, B. Baltimore: Johns Hopkins Press.
CAVALIERI, L. F., ROSOFF, M. & ROSENBERG, B. H. (1956). *J. Amer. Chem. Soc.* **78**, 5239.

CHARGAFF, E. (1955). In *The Nucleic Acids*, 1, 307. Ed. Chargaff, E. & Davidson, J. N. New York: Academic Press.

CHARGAFF, E., CRAMPTON, C. F. & LIPSHITZ, R. (1953). *Nature, Lond.* 172, 289.

COHEN, S. S. (1956). In *Essays in Biochemistry*, p. 77. Ed. Graff, S. New York: John Wiley.

COHEN, S. S. (1957). In *The Chemical Basis of Heredity*, p. 651. Ed. McElroy, W. D. & Glass, B. Baltimore: Johns Hopkins Press.

COLVIN, J. R., SMITH, D. B. & COOK, W. H. (1954). *Chem. Rev.* 54, 687.

DEKKER, C. A. & SCHACHMAN, H. K. (1954). *Proc. Nat. Acad. Sci., Wash.* 40, 894.

DIERMEIER, H. F., DI STEPHANO, H. S. & BASS, A. D. (1955). *J. Pharmacol.* 115, 240.

DOTY, P. (1955). *Proc. Third Internat. Cong. Biochem.*, Brussels, p. 135.

DOTY, P. (1957a). In *The Chemical Basis of Heredity*, p. 391. Ed. McElroy, W. D. & Glass, B. Baltimore: Johns Hopkins Press.

DOTY, P. (1957b). *J. Cell. Comp. Physiol.* 49, suppl. 1, 27.

DOTY, P. & RICE, S. A. (1955). *Biochim. Biophys. Acta*, 16, 446.

DUNN, D. B. & SMITH, J. D. (1954). *Nature, Lond.* 174, 305.

DUNN, D. B. & SMITH, J. D. (1955). *Nature, Lond.* 175, 336.

EPHRUSSI-TAYLOR, H. (1955). *Advances in Virus Research*, 3, 275.

FRAENKEL-CONRAT, H., SINGER, B. A. & WILLIAMS, R. C. (1957). In *The Chemical Basis of Heredity*, p. 501. Ed. McElroy, W. D. & Glass, B. Baltimore: Johns Hopkins Press.

FRIEDKIN, M. & WOOD, IV, H. (1956). *J. Biol. Chem.* 220, 639.

GIERER, A. & SCHRAMM, G. (1956a). *Nature, Lond.* 177, 702.

GIERER, A. & SCHRAMM, G. (1956b). *Z. Naturf.* 11b, 138.

HOGEBOOM, G. H., SCHNEIDER, W. C. & STRIEBICH, M. J. (1952). *J. Biol. Chem.* 196, 111.

HOTCHKISS, R. D. (1955). In *The Nucleic Acids*, 2, 435. Ed. Chargaff, E. & Davidson, J. N. New York: Academic Press.

HOTCHKISS, R. D. (1957). In *The Chemical Basis of Heredity*, p. 321. Ed. McElroy, W. D. & Glass, B. Baltimore: Johns Hopkins Press.

HOTCHKISS, R. D. & MARMUR, J. (1954). *Proc. Nat. Acad. Sci., Wash.* 40, 55.

KAWADE, Y. & WATANABE, I. (1956). *Biochim. Biophys. Acta*, 19, 513.

KAY, E. R. M., SIMMONS, N. S. & DOUNCE, A. L. (1952). *J. Amer. Chem. Soc.* 74, 1724.

LEIDY, G., HAHN, E. & ALEXANDER, H. E. (1953). *J. Exp. Med.* 97, 467.

LEVINTHAL, C. (1956). *Proc. Nat. Acad. Sci., Wash.* 42, 394.

LUCY, J. A. & BUTLER, J. A. V. (1954). *Nature, Lond.* 174, 32.

MIURA, K. & SUZUKI, K. (1956). *Biochim. Biophys. Acta*, 22, 565.

PAHL, H. B., BEISER, S. M. & BENDICH, A. (1957). *Fed. Proc.* 16, 230.

PETERSON, E. A. & SOBER, H. A. (1956). *J. Amer. Chem. Soc.* 78, 751.

PIRIE, N. W. (1938). In *Perspectives in Biochemistry*, p. 11. Ed. Needham, J. & Green, D. E. Cambridge University Press.

PIRIE, N. W. (1952). *New Biol.* 12, 106.

PIRIE, N. W. (1954). *New Biol.* 16, 41.

POUYET, J. & WEILL, G. (1957). *J. Polym. Sci.* 23, 739.

SADRON, C. (1955). *Proc. Third Internat. Cong. Biochem.*, Brussels, p. 125.

SCHACHMAN, H. K. (1957a). In *The Chemical Basis of Heredity*, p. 543. Ed. McElroy, W. D. & Glass, B. Baltimore: Johns Hopkins Press.

SCHACHMAN, H. K. (1957b). *J. Cell. Comp. Physiol.* 49, suppl. 1, 71.

SCHUMAKER, V. N. & SCHACHMAN, H. K. (1957). *Biochim. Biophys. Acta*, 23, 628.

SCHWANDER, H. & SIGNER, R. (1950). *Helv. Chim. Acta*, 33, 1521.

SHERRATT, H. S. A. & THOMAS, A. J. (1953). *J. Gen. Microbiol.* 8, 217.

SHOOTER, K. V. & BUTLER, J. A. V. (1955). *Nature, Lond.* **175**, 500.

SHOOTER, K. V. & BUTLER, J. A. V. (1956). *Trans. Faraday Soc.* **52**, 734.

SHOOTER, K. V., PAIN, R. H. & BUTLER, J. A. V. (1956). *Biochim. Biophys. Acta*, **20**, 497.

SINSHEIMER, R. L. (1954). *J. Biol. Chem.* **208**, 445.

SINSHEIMER, R. L. (1957). *Science*, **125**, 1123.

THOMAS, R. (1954). *Biochim. Biophys. Acta*, **14**, 231.

WECKER, VON E. & SCHÄFER, W. (1957). *Z. Naturf.* **126**, 415.

ZAMENHOF, S. (1956a). In *Essays in Biochemistry*, p. 322. Ed. Graff, S. New York: John Wiley.

ZAMENHOF, S. (1956b). *Progr. Biophys.* **6**, 85.

ZAMENHOF, S. (1957). In *The Chemical Basis of Heredity*, p. 351. Ed. McElroy, W. D. & Glass, B. Baltimore: Johns Hopkins Press.

ZAMENHOF, S., ALEXANDER, H. E. & LEIDY, G. (1953). *J. Exp. Med.* **98**, 373.

ZAMENHOF, S., REINER, B., DE GIOVANNI, R. & RICH, K. (1956). *J. Biol. Chem.* **219**, 165.

SIZE LIMITATIONS GOVERNING THE IN-CORPORATION OF GENETIC MATERIAL IN THE BACTERIAL TRANSFORMATIONS AND OTHER NON-RECIPROCAL RECOMBINATIONS

By ROLLIN D. HOTCHKISS

The Rockefeller Institute for Medical Research,
New York City

In recent years the realm of microbial genetics has provided several instances of non-reciprocal genetic recombinations. By way of definition, one may take a recombination to be non-reciprocal if the two parents make unequal contributions to the net result. In some instances this type of interaction would seem to be the only one possible; as, for example, when a particle of deoxyribonucleate (DNA) interacts with the whole bacterial genome in bacterial transformation (Hotchkiss, 1956) or transduction (Demerec, Blomstrand & Demerec, 1955), or when a fragment of bacterial 'chromosome' enters another cell in bacterial recombination (Wollman, Jacob & Hayes, 1957). Symmetrical exchange between such large and small entities would be expected to yield a recombinant cell plus a recombinant fragment which would have no intrinsic viability and would pass unrecognized. In other cases, where reciprocal interaction would seem to be possible, it only occurs on a statistical basis, and the results show that at any individual event only unidirectional transfer of genetic elements has taken place. Such seems to be the case when two kinds of immature bacteriophage interact within a bacterial host cell, and may also hold for certain recombinations observed in higher forms such as *Aspergillus* (Pritchard, 1955), *Neurospora* (Mitchell, 1957; St Lawrence & Bonner, 1957), and yeast (Winge, 1955), including those designated 'gene conversions' (Lindegren, 1953).

Bacterial transformation offers an interesting example of such a process, since it appears to represent an orderly and predictable interaction between a single growing cell and a single unit of DNA (Hotchkiss, 1957). Two principal classes of hypotheses have been offered to account for the genetic incorporation in transformation: (*a*) those in which a true exchange occurs between the large parental and small infecting units, followed by disposal of the smaller non-viable fragment which is a product of the exchange, and (*b*), those in which the infecting fragment, either by being directly incorporated, or by influencing the replication processes (copy-choice), specifically modifies a particular region of a growing genetic strand, which would otherwise have been laid down as a simple replica of the pre-existing parental structure. The second mechanism was believed to hold for transformation, since the corresponding region with genes from the cell parent seems not to be eliminated, but reappears in some of the progeny of transformed cells, while an incoming fragment bearing linked markers may or may not be totally used (Hotchkiss, 1956). The same mechanism is applicable for all of the genetic systems mentioned above, principally because in all of them a more or less stable 'heterozygote' can be recognized or inferred, temporarily bearing genes from both parents.

Accordingly, a fuller understanding of the events associated with recombination may be informative not only about the structural or formal aspect of genetic chemistry, but may provide valuable information concerning the mechanism of replication of DNA. For if recombination in these systems is the result of some intervention in the normal replication process, the deoxyribonucleate molecules which we can send into the cell in transformation are among the most sensitive and specific reagents which we can bring to bear upon the replication process itself. This article will deal with the accumulating evidences that bacteria undergoing transformation can alter the incoming DNA particle in various subtle ways having both qualitative and quantitative effects. The bearing of these findings upon other genetic systems will be briefly considered.

RECOMBINATION DURING TRANSFORMATION

It is probable that all transformation involves *reassortment* between genomes of donor and recipient cell. Ever since it was demonstrated that neither penicillin resistance nor streptomycin resistance was transferred together with the trait of encapsulation by the DNA of strains having both properties (Hotchkiss, 1951), it has been customary to say that independent action of different markers is 'expected' in transformation. Subsequent work has provided other examples of independent assortment (Austrian, 1953; Hotchkiss & Marmur, 1954). It was chiefly from these findings that the current view developed of DNA molecules with different biological specificity and function existing within the DNA from a given kind of cell, and from this basis came the hopes for differential fractionation or inactivation of biologically active DNA particles.

When there is some reason to believe that two determinants are carried within the same DNA particle, their becoming separated is considered a *recombination*. Just as is the case with reassortment, recombination is recognized by the appearance of progeny different from either parent. It is not possible to say at present whether the two phenomena are different in principle or merge imperceptibly into each other, when slightly linked, nearly independent, markers are involved. The first signs of some sort of recombination in a transformation process were recognized as a potentiation between certain determinants having to do with the efficiency or rate of production of capsular substance (Taylor, 1949; Ephrussi-Taylor, 1951). The chief difficulty with an interpretation involving recombination was that the determinant for full encapsulation, once reconstituted from determinants for partial encapsulation, could not be shown to dissociate again into its 'component parts'. Thus, although the agent determining full encapsulation behaves as a single DNA particle affecting a single cellular character, there is no assurance that the partial determinants which can give rise to it but do not seem to be derivative from it, are contained within it. More recent studies have reported the production of transformants producing two different capsular substances bearing determinants for both of these two antigens in *Haemophilus* (Leidy, Hahn & Alexander, 1953), and in *Pneumococcus* (Austrian & Bernheimer, 1957). In these latter cases the DNA from maximally (doubly) encapsulated strains does give rise to transformants bearing the single capsule types, as well as doubly encapsulated transformants.

In all of these cases genetic linkage was inferred from the qualitative order of frequency with which maximally encapsulated transformants can be recovered. The relation of these determinants to a single biochemical function is suggested, but by no means demonstrated, by their effect upon the single property of encapsulation.

Linkage, and recombination within a linkage group, could be more adequately demonstrated when quantitative selective markers became available. Linkage was demonstrated in this fashion between mannitol utilization and streptomycin resistance in *Pneumococcus* (Hotchkiss & Marmur, 1954). The DNA from doubly marked strains conferred the two properties upon recipient cells far more often than they could have been introduced by chance alone, as for example by a mixture of singly marked DNA's. The same was true when the DNA carried the alternative manifestation of the mannitol character, the factor for non-utilization (absence of specific dehydrogenase), or when the streptomycin marker introduced sensitivity rather than resistance to streptomycin. About one-tenth of the transformants became like the donor in both of these two properties, in either of their manifestations. The remainder of the transformants are recombinants, 40–50 % receiving one of these two markers from the donor strain, and a like number the other.

VARIABLE SIZE OF THE TRANSFORMING UNIT

The picture of transformation which emerged from these results was that a recipient cell, confronted with a particle of DNA from a donor cell, might incorporate and accept genetically a doubly or singly marked portion of that particle. More often it would be a portion carrying one marker. It does not seem likely that the large and small portions are produced by fragmentation or inactivation during isolation and preparation of DNA, since increased exposure to these procedures and isolation of DNA from a wide variety of strains did not lead to change in the linkage properties (Hotchkiss, 1956). What does seem likely is that a more or less intact DNA molecule is absorbed and incorporated chemically, and that only a portion of this molecule survives unchanged to become genetically effective upon the progeny of the transformed cell. The battleground for this struggle between native and introduced DNA determinants may well be the partial heterozygote which is believed to be the first product of interaction between recipient cell and donor DNA. If that struggle is some form of interference with normal replication, then it would seem to involve a substitution of a preformed fragment, sometimes larger, more often smaller, of DNA into a place intended for newly formed material. If we had precise knowledge about the chemical modulations which presumably occur along the poly-nucleotide chains, we might hope to study the intimate metabolic syntheses and exchanges in which the atoms of the gene substance are involved. More available at present are certain subtle genetic modulations which can be identified in the DNA, and the metabolism of these genetic units, i.e. recombination, offers a more convenient current approach to the same interactions.

In 1955, a pneumococcal mutant with high resistance to sulphonamide was discovered by Miss Evans and the author. A significant feature of this system was that although the resistance bred true by cell division, it did not by transformation. The majority of the transformants had lower sulphonamide resistance than the donor, and several lower levels of resistance showed up among them. Genetic analysis, which has been briefly reported (Hotchkiss & Evans, 1957), indicates that the highly resistant mutant contains three different determinants, presumably all introduced at a single mutational event. This can be described as a fine structure in the DNA, and symbolized as *adb*. Three unit types, *a*, *d*, and *b*, could be recognized among the transformants, and three complex types, *ad*, *db* and the original *adb*, all having characteristic degrees of drug resistance. When these strains were used in turn as sources of DNA for further transformation, the determinants *a*, *d*, and *b* reappeared as units just as expected from the complex types, or could as donors be introduced as simple units into sensitive cells, or be reassembled into complexes by crossing with different unit types.

The fine structure indicated seems, therefore, similar to those recognized somewhat earlier by Benzer (1955) in bacteriophage and by Demerec *et al.* (1955) in bacterial determinants conveyed by virus transduction. In the present case, however, the fine structure is detectable in isolated DNA subject to chemical study, degradation, and the like. Furthermore, the pneumococcal subtypes can each be recognized by their characteristic phenotypic levels of sulphonamide resistance, as well as by their genetic capability to recombine and restore a lacking property. Arguments have been presented (Hotchkiss & Evans, 1957) which tend to indicate that the mutant and transformant subtypes all differ from the sensitive wild-type strain in the properties of a single-enzyme system using para-aminobenzoate for the synthesis of folic acid.

The frequencies with which these complex and unitary determinants are introduced through transformation indicates that there are indeed size limitations on the DNA incorporated genetically. For example, in one experiment with *adb* as donor of DNA to sensitive type cells, approximately 5% of the cells were transformed in sulphonamide resistance. Only 0·4% of these transformants were the triplex *adb* like the donor, the duplex types *ad* and *db* made up 20%, and the three simple unit types constituted the other four-fifths of the transformants. Very similar proportions were recovered when many different preparations of DNA from this mutant strain or its derivatives acted upon the sensitive strain from which the mutant originated. When one of the unit subtypes is used as recipient, the appropriate duplexes are produced frequently, since only a single new determinant need be added to produce them in this case.

These results go far to document the inference, drawn from the mannitol-streptomycin case, that large units of DNA cannot often be incorporated genetically. For in the earlier two-marker system, single and double markers might have been introduced by fragments of similar size taken from different regions, and it was principally from the analysis of transformant clones that one could infer that only a fragment of the entering DNA was usually utilized. In the 'disseminative' transformation of the three-marker sulphonamide resistance, it is the simpler determinants which are most often introduced from the complex DNA's, and each simpler type clearly contains a smaller amount of *mutated* DNA than the complexes (and inspires a lower drug-resistance), whether one is comparing the units with the duplexes, or the duplexes with the triplex. For the simple units are *contained in*, and can be obtained from, the complexes, but they *do not contain* the more complex determinants, and can only produce them when recombined with the appropriate other units. Therefore, one is entitled to say that in this system, the determinants are of different sizes and are introduced with a frequency varying inversely according to their size. The largest group, *adb*, is only rarely incorporated genetically intact.

The reaction of cells with DNA probably involves in its initial phases entire DNA molecules, since it is rapid, occurs best and is best inhibited with intact undenatured DNA (Hotchkiss, 1953–4, 1954), is one-hit in its expression (Ravin, 1954; Hotchkiss, 1957), follows a clear-cut kinetic course (Thomas, 1955; Fox & Hotchkiss, 1957) and appears to result in either negligible or else sharply defined, permanent uptake of ^{32}P-DNA (Goodgal & Herriott, 1957; Lerman & Tolmach, 1957; Fox, 1957). The results of Schaeffer presented at this symposium point in the same direction. It seems, therefore, that transforming DNA is taken up as whole molecules and chemically is completely utilized for cellular DNA, but biologically speaking, only fragments, usually small, survive to produce a genetic change.

In another investigation with Mr S. Lacks of this laboratory, it has recently been found that the DNA determining maltose hydrolysis by *Pneumococcus* can be damaged in different closely linked sites by the mutagenic action of ultra-violet light. Recombination tests between the maltose-negative mutants produced reveal a fine structure with five very small mutant regions, two larger regions overlapping several of these smaller ones, and one still larger mutant region or deletion. The map deduced from qualitative abilities to recombine is supported amply by the frequency map from quantitative comparisons of recombination rate corrected for overall transformation rate. What is important to note here is that, in a general way, the larger deficiencies show a lower frequency of 'repair' than the smaller ones, when being transformed back to maltose utilization by one and the same wild-type DNA preparation. Here again, large determinant regions seem to be introduced less often than small ones from the same DNA. Lacks has also demonstrated that the larger regions are more quickly activated than the smaller ones when this DNA is partially denatured by heat, another form of demonstration that they are of greater size.

NATURE OF THE DNA DISSEMINATION DURING RECOMBINATION

The selection, and preservation or disruption, of such submolecular DNA fragments would presumably be decided by metabolic events within the individual cells undergoing transformation. Although there is a tendency to assume randomness even when such intramolecular events are concerned, there is actually no *a priori* reason for supposing that the decision made by a number of cells would necessarily represent statistically random choices for the sites of breakage along the molecule. What is actually meant, of course, is random breakage at some assumed class of linkages, such as the $3'$- or the $5'$-phospho-ester linkages or at hydrogen-binding sites, or more generally perhaps, random breakage among all the bonds of a *particular grade of lability*. This, it can be seen, is far from a bold and precise counsel of wisdom, being rather an adaptable statement of ignorance. The best sign that some such molecular factors prevail, be they random or highly specific, is the reproducibility of the linkage patterns in general when cells are transformed under different conditions, and the dependence of these patterns on the state of the DNA (current investigations in the author's laboratory). If we could perceive with some clarity the factors determining sites of breakage and fusion (or synapsis and copy-error if that is the mechanism) within DNA molecules undergoing recombination, we would probably learn a great deal more about the processes of DNA synthesis and replication.

The remainder of this article will be devoted to an exploration of the limitations placed on random recombination by a size limit governing the fragments that can be accommodated. The considerations which follow are believed to be applicable to all recombinations which occur at the submolecular level. These are probably nearly all recognizable as recombinations which as individual events are non-reciprocal: the transformations, transductions, phage and bacterial recombinations, and short-range crossings over or gene conversions, such as those reported in the moulds and fungi mentioned above. Such generalization is offered because it seems to reduce some of the apparent contradictions and complications that have recently appeared in the study of such unidirectional recombinations.

SIZE-LIMITED NON-RECIPROCAL RECOMBINATION

A linear array will have at least one dimension, length, and—if it is differentiated along this dimension—a direction may also be defined. Thus, each fragment may have a size and a 'beginning' and 'end'. Size may be measured in units of equal recombination frequency, so that one escapes the necessity of assuming the random or specific nature of breakage at particular linkages. In Fig. 1, the unit, $\times$, of recombination frequency, is considered to be large enough to include the entirety of the largest of the three marker regions A, B, and C. The total recombining length, N, the intermarker distances l_1 and l_2, etc., are given in terms of *the number of sites for breakage* (number of units plus one).

Consideration of the specific case presented in Fig. 1 will reveal that assigning a maximum size, m, to the DNA which can be transferred (or imprinted), sharply modifies the random expectation. Simply taking N to be large, one would expect very few opportunities for ABC to appear. But a size limitation greatly diminishes the number of AB and BC recombinants formed, since many of the smaller fragments (represented as those ending at o in the figure) would be contained between the two

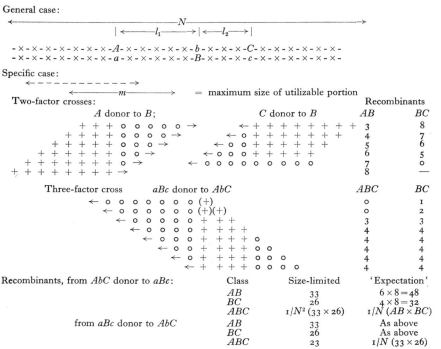

NOTE: Of all fragments terminating at →, those having their other terminus at o do not lead to recombinants, while those at + do so.

Fig. 1

markers and not be detected. Thus, the 'distances' AB and BC are minimized and one expects too few ABC recombinants if one supposes them to be the resultant of two rare events like those leading to AB and BC. Actually, for the case assumed, the number of ABC-type (23) approaches that of the more rare simple type (26), a finding which would be classed as high 'negative interference' in a standard analysis.

It will be obvious, also, that the 'distance', AC, will seem smaller than the sum of $AB + BC$, since it will bracket an even greater number of undetected captures than can the smaller intervals. Conversely, shorter intervals may be considered exaggerated or 'expanded' if the longer distances are thought more reliable. The basic difficulty is that one does not actually know for any of these recombinations either N, the potential number of crossing sites and occasions, nor $1/N^2$, the probability for any given double insertion, nor in fact the manner in which the crossing occurs. Therefore, the frequencies themselves must be somewhat arbitrarily calculated in relation to total particles examined, for example, and it is not possible to relate them meaningfully to absolute physical distances with any confidence, while calculating the product of two frequencies compounds the error from the value taken for N. The 'expected' values in Fig. 1 are based upon the assumption of a single crossover and of the chosen distances, adjusting the frequencies (eight events at each site) to suit the order of magnitudes imposed by the size-limited dissemination model (eight fragments originating at any one point).

For the general case of Fig. 1, recombination over an actual distance l between markers is calculated as

$$l/2N^2 (2m - l - 1)$$

for unidirectional transfer, or twice as great for systems (bacteriophage, fungi) in which both parents can serve, though not simultaneously, as donor. It will be appreciated that this is approximately proportional to l (like the simple relation l/N usually inferred) when m is large relative to l. The three-factor recombinant ABC would appear at a frequency

$$l/2N^2 [m^2 - (m - l_1)^2 - (m - l_2)^2 - (l_1 + l_2 - m)].$$

These expressions are valid only when $m > l + 1$, that is, when the markers appear at least slightly linked. For the special case assumed, l_1 was 6, $l_2 = 4$, $m = 9$, and the frequencies were calculated per N^2 recombinants.

Failure of additivity, expansion of the scale at short distances, and high 'negative interference' have frequently been observed in mapping of genetic fine structure in phage (Benzer, 1955; Chase & Doermann, 1958) and fungi (Pritchard, 1955; St Lawrence & Bonner, 1957). These effects may be qualitatively rationalized if the size-limited disseminative recombination, indicated by direct analysis as occurring in the DNA transformations, be assumed to hold for all of these non-reciprocal recombinations. It should be emphasized that this picture does not imply direct correspondence between the sizes of recombination units in the different systems, since for different systems the limitations may well become restricting at different levels. However, it does tend to carry the implication of some maximum size beyond which recombination is not between DNA but between DNA-carrying chromosomes or their equivalent. It would seem likely, but not altogether certain, that this boundary provides the dichotomy between the non-reciprocal and the reciprocal recombination. Nevertheless, there is no clear sign of a dichotomy between small and large regions as to linkage behaviour.

More particularly, the hypothesis, and the data from which it derives, make it more difficult to judge the relation of markers that do not appear to be linked. Markers sufficiently separated not to show these complex interactions may still show classical linkage, as in *Escherichia coli*, the fungi and *Drosophila*. According to the view presented here, in transforming DNA and phage, the non-reciprocal recombinations would presumably reach a maximum between markers separated beyond a certain limit, and beyond this, there would seem to be no linkage. Therefore, the assumption to which reference was previously made, that many kinds of DNA molecules, bearing different specificities, exist independently in these preparations, still remains unproved.

In the above we have done something other than elaborate a hypothesis of double crossover. The distinctive specifications should perhaps be summarized: (1) disseminative recombination is applied to DNA molecules rather than to chromosomes; (2) one particle may be considered the donor, another the recipient; (3) the transferred or imprinted fragment has a limited size, a beginning, an end, and a direction; (4) the size, and sites of insertion or copying, are very likely governed by biochemical events rather than primarily by physical positioning; and (5) it is likely that exchange requires close compatibility along the whole synapsed interval.

If size limitations provide one restriction upon recombinations, another specific restriction on randomness may come from the directional polarity which DNA strands possess. The single strands of DNA bear a polarity due to -3'-phospho-5'-deoxyribo-3'-phospho-5'-deoxyribo- linkages. This may be of importance for the ultimate theories of replication and recombination, since the events at respective ends of an incorporated fragment may be

$$\text{terminal-5'-PO}_4 + \text{fragment} \rightarrow \text{-5'-3'-diphosphate, or}$$
$$\text{terminal-5'-OH} + \text{PO}_4\text{-fragment} \rightarrow \text{-5'-3'-diphosphate}$$

at one end and

$$\text{terminal-3'-PO}_4 + \text{fragment} \rightarrow \text{-3'-5'-diphosphate, or}$$
$$\text{terminal-3'-OH} + \text{PO}_4\text{-fragment} \rightarrow \text{-3'-5'-diphosphate}$$

at the other. Phosphodiesterases and polynucleotide-synthesizing systems are already known which can distinguish with sharp specificity these two ends from each other. Oriented in time and space, such enzymes might specify the sites for beginning or ending an insertion or copying. Theories of DNA replication should take into account the enormous stability one single helix running in one direction could have towards an enzyme which may be actively engaged in exchanging base residues at the end of a chain which is being synthesized in close proximity along an opposite direction. If, instead, double helices are the active form of the DNA, similar considerations may apply to the more sophisticated directional properties presumably conferred upon them by modulations in base distribution along the paired helices.

Taken in their most general form the results of recent transformations suggest that we are observing signs of processes that restrict the randomness and control the place of recombination in DNA molecules. We may be nearing the time at which we can more closely specify the nature of the biochemical reactions themselves.

REFERENCES

AUSTRIAN, R. (1953). *J. Exp. Med.* **98**, 35.

AUSTRIAN, R. & BERNHEIMER, H. (1957). In *The Chemical Basis of Heredity*, p. 346. Ed. McElroy, W. D. & Glass, B. Baltimore: Johns Hopkins Press; and unpublished results.

BENZER, S. (1955). *Proc. Nat. Acad. Sci., Wash.* **41**, 344.

CHASE, M. & DOERMANN, A. H. (1958). (in the Press).

DEMEREC, M., BLOMSTRAND, I. & DEMEREC, Z. E. (1955). *Proc. Nat. Acad. Sci., Wash.* **41**, 359.

EPHRUSSI-TAYLOR, H. (1951). *Exp. Cell. Res.* **2**, 589.

FOX, M. S. (1957). *Biochim. Biophys. Acta* (in the Press).

FOX, M. S. & HOTCHKISS, R. D. (1957). *Nature, Lond.* **179**, 1322.

GOODGAL, S. & HERRIOTT, R. M. (1957). In *The Chemical Basis of Heredity*, p. 336. Ed. McElroy, W. D. & Glass, B. Baltimore: Johns Hopkins Press.

HOTCHKISS, R. D. (1951). *Cold Spr. Harb. Symp. Quant. Biol.* **16**, 457.

HOTCHKISS, R. D. (1953-4). *Harvey Lect.* **49**, 124.

HOTCHKISS, R. D. (1954). *Proc. Nat. Acad. Sci., Wash.* **40**, 49.

HOTCHKISS, R. D. (1956). In *Enzymes: Units of Structure and Function*, p. 119. Ed. Gaebler, O. H. New York: Academic Press.

HOTCHKISS, R. D. (1957). In *The Chemical Basis of Heredity*, p. 321. Ed. McElroy, W. D. & Glass, B. Baltimore: Johns Hopkins Press.

HOTCHKISS, R. D. & EVANS, A. H. (1957). In *Symposium on Drug Resistance in Micro-organisms*, Ciba Foundation, London, p. 183.

HOTCHKISS, R. D. & MARMUR, J. (1954). *Proc. Nat. Acad. Sci., Wash.* **40**, 55.

LEIDY, G., HAHN, E. & ALEXANDER, H. E. (1953). *J. Exp. Med.* **97**, 467.

LERMAN, L. S. & TOLMACH, L. E. (1957). *Biochim. Biophys. Acta* (in the Press).

LINDEGREN, C. C. (1953). *J. Gen.* **51**, 625.

MITCHELL, H. K. (1957). In *The Chemical Basis of Heredity*, p. 94. Ed. McElroy, W. D. & Glass, B. Baltimore: Johns Hopkins Press.

PRITCHARD, R. H. (1955). *Heredity*, **9**, 343.

RAVIN, A. W. (1954). *Exp. Cell. Res.* **7**, 58.

ST LAWRENCE, P. & BONNER, D. M. (1957). In *The Chemical Basis of Heredity*, p. 114. Ed. McElroy, W. D. & Glass B. Baltimore: Johns Hopkins Press.

TAYLOR, H. E. (1949). *J. Exp. Med.* **89**, 399.

THOMAS, R. (1955). *Biochim. Biophys. Acta*, **18**, 467.

WINGE, O. (1955). *C.R. Lab. Carlsberg*, **25**, 341.

WOLLMAN, E. L., JACOB, F. & HAYES, W. (1957). *Cold Spr. Harb. Symp. Quant. Biol.* **21**, 141.

INTERSPECIFIC REACTIONS IN BACTERIAL TRANSFORMATION

By PIERRE SCHAEFFER

Service de Physiologie microbienne, Institut Pasteur, Paris

Discovered by Griffith (1928) in *Pneumococcus*, bacterial transformation remained for many years unique to this species. Transformable strains were eventually found in other species, for example *Haemophilus influenzae* (Alexander & Leidy, 1950). But for a quarter of a century, transformation remained an intraspecific phenomenon, the donor of DNA either being derived by mutation from the receptor strain, or being another strain of the same species. Recently, interspecific transformations were shown to occur within the genus *Haemophilus* (Schaeffer & Ritz, 1955; Alexander & Leidy, 1955), and it is these which are the subject of the present paper.

A feature common to all interspecific transformations is their low frequency as compared with the intraspecific kind. That the ability of a recipient culture to become transformed, i.e. its competence, is a transient property, is agreed upon by many authors (review in Hotchkiss, 1957). But differences in competence cannot be invoked in interspecific transformation, as all frequencies are determined on aliquots of the same recipient culture, simultaneously exposed to the various DNA's. The state of competence of the culture is found to affect equally the frequency of both inter- and intraspecific transformation, the ratio of the two frequencies being nearly constant from one experiment to the other. This ratio is also found to be independent of such variables as DNA concentration, duration of contact between DNA and recipient bacteria, and time allowed for phenotypic expression (Schaeffer, 1956). It does not vary significantly from one DNA preparation to the next, and in fact, it has not been possible to alter it in any way.

In order to expose the questions raised by such unequal transformation frequencies, it might be useful to start with a formal scheme of the succession of events that must be at work in transformation (Hotchkiss, 1956; see also Demerec & Demerec (1956), as events following penetration are considered to be common to transformation and transduction). These events, in order, are (1) *adsorption* of the transforming DNA molecule to a site on the surface of the bacterium; (2) its *penetration* inside the cell; (3) its *specific pairing*, or *synapsis*, with a chromosomal structure in one of the nuclei (whether this structure is DNA, or some template of a different nature, a theoretically important question, is a minor point in the present study); (4) *incorporation* of the genetic factor brought into the cell with the transforming molecule, into a chromosomal DNA structure; this step ensures duplication of the factor at each division cycle. Material incorporation seems excluded, however, and use of the word *integration* to designate this step is therefore preferable; (5) *segregation* of the transformed nucleus in one or two mitotic divisions. The whole process eventually leads to a transformant, in which a new character is phenotypically expressed; *expression*, however, a physiological phenomenon, is unique in not being assigned a defined location in the succession of events. Its onset might immediately follow penetration, or necessitate integration, or even be postponed until after segregation has taken place, depending on dominance relationships.

Let us now examine how these formal steps compare with those that we are able to recognize. If competent bacteria are exposed to ^{32}P-labelled DNA, they pick up radioactivity, and retain it despite further washings and DNase treatment (Goodgal & Herriott, 1957; Fox, 1957). Furthermore, if these cells are lysed by gentle treatment, transforming activity is recovered from the lysate, in amount accounting for the radioactivity retained (Goodgal & Herriott, 1957). It is not quite clear whether this retention of DNA, in a DNase resistant form, should be taken to mean actual penetration inside the cell wall. It seems wise, therefore, as long as adsorption and penetration are not experimentally recognized as separate phenomena, to speak only of fixation, retention, or, if it must be, 'penetration' of the exogenous DNA molecule, when referring to the observed facts.

Step 3, pairing, is still awaiting direct cytological demonstration; it stands as a logically necessary prerequisite to integration, and is believed to demand structural identity, or precise complementarity, of the two partners involved.

The mechanism of step 4, integration, has been discussed by Hotchkiss (1956) and by Demerec & Demerec (1956), and is likely to be a copy-error, or a switch of the DNA replicating mechanism from the endogenous to the exogenous model, and back, so that a short segment of the transforming DNA molecule is eventually built in the new copy. By using donor strains with several genetic markers, or by studying the frequency of intra- and interspecific transformations, as is done below, information can be gained on the relative length of this segment; nevertheless, study of integration, in transformation, is still of an indirect nature.

The last step, segregation, is clearly amenable to direct study, at least in Hotchkiss's skilful hands (Hotchkiss, 1956).

Going back now to interspecific transformations, one may wonder (since competence and expression have already been excluded as candidates), which of the steps involved in transformation is responsible for the low frequencies observed. This is the subject-matter of the first part of this study, in which it is shown that the experimental data can be accounted for in terms of pairing and integration, and that differences in fixation are indeed not involved.

But interspecific reactions, as referred to in this paper, are not restricted to interspecific transformation proper; they include also inhibition reactions in which the receptor strain, and the strains supplying the transforming and the inhibitory DNA's, do not belong in the same species. The second part of this paper deals, therefore, with inhibition of transformation by DNA. An explanation is given for the fact (Hotchkiss, 1957), that despite the known competitive nature of inhibition (Hotchkiss, 1954; Alexander, Leidy & Hahn, 1954), no inhibition occurs when the concentration of transforming DNA is low. The inhibition data presented can also be made to mean, that practically every bacterium is competent in a fully competent culture of *Haemophilus influenzae*. The site of the competition occurring in inhibition has been determined by using radioactive transforming DNA, and found to be located at the cell surface; this location is consistent with the results obtained in the study of interspecific inhibition reactions. Finally, the nature and the specificity of the surface receptor are discussed.

I. INTERSPECIFIC TRANSFORMATION
General information

(1) Three strains of bacteria have been used in this work; they are designated by the symbols Rd, Fid and A61. The necessary information about them is presented in Table 1.

(2) Only one genetic marker has been used in all the transformation experiments to be described, namely resistance to high concentrations of streptomycin. The marker was chosen for its rare occurrence, its acquisition through one-step mutation, and its easy, quantitative selection. Correct genetic symbols for streptomycin sensitivity and resistance are Sm^s and Sm^r, respectively; in this paper, however, for the sake of simplicity, Sm^s will be omitted and Sm^r reduced to Sm.

(3) Deoxyribonucleic acid is written DNA, and deoxyribonuclease DNase.

(4) Transformation of, for example, recipient strain Rd by DNA from the mutant RdSm, is written Rd × (RdSm), parentheses indicating DNA preparation.

(5) Designations such as FidSm are confined to spontaneous mutants alone. A resistant clone formed in transformation Fid × (FidSm) is named FidSmFid. Similarly, it can be seen from its designation that RdSmA61 is a transformant formed in the reaction Rd × (A61Sm). In a similar fashion, transformants formed in the reaction Fid × (RdSmFid) are named FidSmFid, Rd. With this system, therefore, the first strain symbol refers to the nature of the strain and those following the Sm symbol enumerate,

Table 1. *Information on the strains employed*

Strain symbol	Species	Ability to synthesize		Transformability by DNA from its own mutants	Observations
		Haematin	DPN		
Rd	*Haemophilus influenzae*	o	o	+	Rough mutant from wild type of serotype d; obtained from Dr H. E. Alexander
Fid	*H. parainfluenzae*	+	o	+	Isolated from human throat; obtained from Dr H. E. Alexander
A61	*H. suis?*	+	o	o	Kept in collection, labelled *H. parainfluenzae*; origin unknown; behaves in transformation experiments like those strains of swine origin that Leidy, Hahn & Alexander (1956) proposed to call *H. parasuis*

in order, the strains in which the Sm determinant first appeared and was passed along. Justification for this system is to be found in the fact that the biochemical character, upon which the distinction between the species depends (ability to synthesize haematin), is not transferred in transformation (Schaeffer, 1956).

(6) Hotchkiss's rigorous definition of the frequency of transformation has been followed (Hotchkiss, 1957); it is defined as the ratio of the number of transformants, counted after expression is completed and before their multiplication starts, to the number of viable cells present at the end of a short ($\leqslant$ 10 min.) exposure to DNA, i.e. at time of addition of DNase.

Spontaneous mutants as source of DNA

By combining the data published by Schaeffer (1956) and by Leidy, Hahn & Alexander (1956), the initial observations on interspecific transformation can be summarized in the following statement: while in the homospecific reaction Rd × (RdSm), the frequency of transformation is of the order of 1×10^{-3}, in the heterospecific reactions A[Rd × (FidSm)] and B[Rd × (A61Sm)], the frequencies are close to 3×10^{-6} and 1×10^{-7}, respectively. The homospecific reaction is therefore 300 times more frequent than reaction A and 10,000 times more frequent than reaction B. Our purpose is to find out the meaning of these facts.

As already mentioned, the answer is not to be found in differences in the competence of the recipient cultures, or in unequal rates of expression of the transformants (Schaeffer, 1956). Before any genetic interpretation is attempted, however, the observed transforming abilities of the three DNA's employed must be shown to be intrinsic properties of the DNA's themselves, rather than artefacts introduced with their extraction. It is indeed conceivable that the same extraction procedure might lead to fully active preparations when applied to Rd, and to partly inactivated ones when applied to the other two strains; such would be the case, for instance, if some DNase activity was present in the lysates from Fid and A61.

This possibility is ruled out by studying the reactions Fid × (FidSm) and Fid × (RdSm), in which Leidy et al. (1956) find the frequencies to be, in one experiment, $6·0 \times 10^{-5}$ and $2·7 \times 10^{-7}$, respectively. The preparation of DNA from FidSm is 200 times more active than the one from RdSm, when Fid is taken as the receptor strain. In *Haemophilus*, therefore, whatever the species of the recipient strain, high frequencies of transformation are obtained in intraspecific reactions, while interspecific ones, when observed, are characterized by their much lower frequencies. Our own data, reported in Table 2, confirm this conclusion. The suspicion of a partial inactivation of some DNA preparations being eliminated, the observed properties of all preparations can safely be ascribed to the DNA's themselves, i.e. to their specific structures.

Heterospecific transformants as source of DNA

The question asked now is whether the low frequency of an interspecific transformation is due to the heterospecific origin of the Sm determinant itself, or of the DNA molecule as a whole. In other words, if X and Y are strains of two distinct species, will DNA from XSmY behave in transformation like DNA from YSm or like DNA from XSm?

The results of experiments of this type, where the transforming activities of DNA's from spontaneous mutants and from homo- and heterospecific transformants are compared, are given in Table 2. Most of these results duplicate those of Leidy et al. (1956), with which they are in good agreement. The following comments can be made. The intraspecific reactions (nos. 1–4, 10 and 11, in both series A and B) have high frequencies of transformation, as compared with interspecific ones; they form a remarkably homogeneous group, with their frequencies not significantly different from each other in the same experiment, and it seems to be of no importance in this group whether the Sm determinant originated in one or the other species. This is in contrast with the group of interspecific reactions, where reaction no. 7 yields roughly 3 times as many transformants as reaction no. 6 in series A, and 16 times as many in series B.

Table 2. *Frequency of the transformation of a given receptor strain by DNA's from various homo- and heterospecific donors* (from Schaeffer, P., *Ann. Inst. Pasteur*, in the Press). *Strain symbols as explained in the text*

Receptor strain	Donor strain	Reaction no.	Number of transformants in 10^8 bacteria	
			Expt. 1	Expt. 2
Rd	RdSm	A1	$3\cdot1 \times 10^5$	—
Rd	RdSmRd	A2	$3\cdot6 \times 10^5$	$1\cdot4 \times 10^5$
Rd	RdSmFid	A3	$4\cdot2 \times 10^5$	$1\cdot6 \times 10^5$
Rd	RdSmA61	A4	$3\cdot3 \times 10^5$	$1\cdot3 \times 10^5$
Rd	FidSm	A5	$9\cdot6 \times 10^2$	—
Rd	FidSmFid	A6	$8\cdot8 \times 10^2$	$3\cdot6 \times 10^2$
Rd	FidSmRd	A7	$3\cdot0 \times 10^3$	$9\cdot6 \times 10^2$
Rd	FidSmA61	A8	$4\cdot5 \times 10^2$	$1\cdot5 \times 10^2$
Rd	A61Sm	A9	$3\cdot3 \times 10^1$	—
Rd	RdSmRd, Fid	A10	—	$1\cdot0 \times 10^5$
Rd	RdSmFid, Rd	A11	—	$1\cdot8 \times 10^5$
Rd	FidSmFid, Rd	A12	—	$3\cdot6 \times 10^2$
Rd	FidSmRd, Fid	A13	—	$2\cdot0 \times 10^3$
Fid	FidSm	B1	$3\cdot1 \times 10^4$	$2\cdot6 \times 10^3$
Fid	FidSmFid	B2	$6\cdot6 \times 10^4$	$4\cdot8 \times 10^3$
Fid	FidSmRd	B3	$3\cdot4 \times 10^4$	$2\cdot4 \times 10^3$
Fid	FidSmA61	B4	$4\cdot5 \times 10^4$	$3\cdot6 \times 10^3$
Fid	RdSm	B5	—	—
Fid	RdSmRd	B6	$1\cdot3 \times 10^2$	c. 10
Fid	RdSmFid	B7	$2\cdot0 \times 10^3$	$1\cdot7 \times 10^2$
Fid	RdSmA61	B8	$2\cdot5 \times 10^2$	c. 10
Fid	A61Sm	B9	$6\cdot6 \times 10^1$	c. 1
Fid	FidSmFid, Rd	B10	$3\cdot4 \times 10^4$	—
Fid	FidSmRd, Fid	B11	$5\cdot0 \times 10^4$	—
Fid	RdSmRd, Fid	B12	$2\cdot1 \times 10^2$	—
Fid	RdSmFid, Rd	B13	$2\cdot3 \times 10^3$	—

The only clear-cut answer to the question asked at the beginning of this section is that DNA from XSmY does not behave at all like DNA from YSm. Whether it behaves like DNA from XSm, however, depends on whether it is tested on X or on Y. If it be asked whether the species in which the mutation appeared is of importance for the subsequent transforming abilities of a DNA, one is led to contradictory answers, according to whether intra- or interspecific reactions are being considered. This asymmetrical situation seems to be reasonably accounted for in the interpretation now presented.

The imperfect-pairing hypothesis

According to the scheme of the steps involved in transformation, developed in the introductory part of this paper, the low frequency of interspecific transformation could result from obstacles to adsorption and penetration as well as to pairing. The study of DNA fixation will be presented in a separate section. Our present purpose is to show that imperfect pairing in interspecific transformation accounts for the observed facts.

Let us consider, in a recipient bacterium, the DNA structure carrying the Sm *locus*, or rather that part of the structure (AA' in Fig. 1), within which a copy-error, when started, may encompass the Sm *locus*. In intra-specific reactions, the pairing is good all along the AA' region, and the probability of a copy-error including the *locus* (i.e. the frequency of trans-formation) is the highest. The resulting transformants seem to be identical with the original spontaneous mutant.

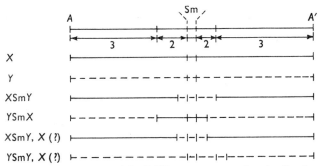

Fig. 1. Schematic representation of the DNA structure carrying the Sm determinant in two species (X and Y) and in transformants resulting from interspecific transformations (from Schaeffer, P., *Ann. Inst. Pasteur*, in the Press). The three zones mentioned in the text are shown; strain symbols as explained in the text.

In interspecific reactions, structural differences in the AA' region between the DNA's of the two species act as obstacles to pairing and there-by reduce the transformation frequency. Unlikely as it is, once an inter-specific transformation of species X by DNA from species Y has occurred, however, *most* of the obstacles to pairing with X must have been eliminated in the XSmY transformants, as is shown by the fact that DNA from XSmY is generally found to be just as efficient in transforming X as DNA from XSmX. That, in the DNA of XSmY, *some* of the Y-specific structures, which behave as obstacles in pairing with X, must have remained, however, is indicated by the fact that higher frequencies are obtained in the reaction $Y \times (X$Sm$Y)$ than in the reaction $Y \times (X$Sm$X)$.

Three zones must therefore be recognized in the AA' region in a transformed bacterium: the Sm-*locus* itself (zone 1), the segments on both sides of the *locus*, inherited together with it (zone 2), and the extreme segments unaffected by this particular copy-error (zone 3). The fact that interspecific reactions are more sensitive than intraspecific ones in detecting the existence of region 2, is taken to proceed from region 2 being smaller than 3. In the reaction $X \times (X\text{Sm}\,Y)$, so many errors encompassing the *locus* are initiated in the long, homospecific region 3, that a few, failing to arise in the short, heterospecific region 2, go unnoticed; in the reaction $Y \times (X\text{Sm}\,Y)$, the few errors initiated in the short, homospecific region 2 become conspicuous, as even less are contributed by the long heterospecific region 3.

The proposed interpretation seems therefore consistent so far with the experimental results; if correct, it means that only a precise pairing, such as has been secured in homospecific regions, provides the opportunity for a copy-error to be initiated, but that, once initiated, the error can extend to heterospecific regions.

As shown in Table 2 (reactions nos. A and B 10–13) the four possible second-order recombinant DNA's have also been prepared and tested on both species. As these DNA's resulted from two successive transformation reactions, their testing was intended to decide whether the segments 'transferred' in these reactions overlapped exactly or not. The results were of doubtful significance, showing only that more was being asked of the method than it is able to answer.

The fixation of heterospecific DNA by the competent bacteria

Implicit in the imperfect-pairing hypothesis is the assumption that DNA adsorption and penetration into the competent bacterium is independent of its origin and specific structure. Data on this point are needed, however, to check upon the hypothesis.

In their study of the fixation of homospecific transforming DNA by Rd cells, Goodgal & Herriott (1957) determined its molecular weight to be 15 million. They expressed their results in terms of number of DNA molecules retained, after washing, by the treated bacterial population for each bacterium transformed. This number was found to be independent of the concentration of DNA and of the frequency of transformation, and to be equal to 120. This does not mean, however, that the number is a constant, as it was later shown by Fox (1957) to vary from one DNA preparation to another, and for a given radioactive preparation, with its age; it seems to depend on the 'quality' of the DNA at the time of its use. Fox's data show a range of variation of one hundredfold: 80–7200 molecules taken up per bacterium transformed, assuming a molecular weight of 15 million. (Fox's work on *Pneumococcus*.)

In order to see whether the retention by competent bacteria of the various DNA's here employed is the same or not, experiments were made by transforming Rd with highly purified preparations of radioactive DNA's from RdSm, FidSm and A61Sm. Partial results have already been published (Schaeffer, 1957*a*). In each individual experiment, the amount of DNA retained was found to be the same, within a factor of less than 2, irrespective of the species used as the donor (Table 3, second column from the right).

Table 3. *Retention of* ^{32}P *DNA by competent Rd bacteria in intra- and interspecific transformations* (from Schaeffer, P., *Ann. Inst. Pasteur*, in the Press)

Experiment	DNA preparation	Specific activity at time of preparation in μc./mg.	Age of preparation (days)	Frequency of transformation	mμg. of DNA retained by DNase-treated bacteria from 1 ml. of culture	DNA molecules (MW = 15 × 10⁶) retained per bacterium transformed
1	RdSm no. 3	234	10	1/400	3·0	210
	FidSm no. 2	312	4	1/120,000	6·0	125,000
2	RdSm no. 5	150	16	1/250	3·9	120
	FidSm no. 3	390	16	1/50,000	2·7	17,000
	A61Sm no. 2	680	6	1/900,000	2·3	260,000
3	RdSm no. 5	150	24	—	8·6	53
	FidSm no. 3	390	24	—	6·7	14,000
	A61Sm no. 2	680	14	—	5·6	2,000,000

In order to express the results in terms of molecules, the molecular weight of 15 million was assumed to be common to all three DNA's. The results in Table 3 are likely to be distorted to some extent by the unpredictable variations in the 'quality' of the DNA preparations employed. The data strongly suggest, however, that such variations cannot affect the validity of the following conclusion: DNA's produced by the strains Fid and A61 are taken up by competent Rd cultures, to the same extent as DNA produced by strain Rd itself. Differences in adsorption and penetration cannot, therefore, be invoked to account for the low frequency of the interspecific transformations studied here, and the imperfect-pairing hypothesis seems all the more valid. From the few data presented, the 'penetration' mechanism, whatever its nature, does not seem to be species-specific.

II. THE INHIBITION OF TRANSFORMATION BY HOMO- AND HETEROSPECIFIC DNA'S

The competitive nature of inhibition

Transformation can be inhibited by addition of DNA not carrying the factor being scored. When both DNA's, transforming and inhibitory, are added simultaneously, the inhibition can be shown to be of the competitive type. Inhibition is also observed when large concentrations of calf-thymus DNA are used (Alexander, Leidy & Hahn, 1954; Hotchkiss, 1954). Recently, Hotchkiss has confirmed the competitive nature of inhibition, but found at the same time that no inhibition occurred when low concentrations of transforming DNA were used (Hotchkiss, 1957). No explanation was offered for these seemingly contradictory observations.

The absence of inhibition when low concentrations of transforming DNA are used, has been observed as well with *Haemophilus influenzae*. For example in the reaction Rd × (RdSm), inhibited by simultaneous addition of Rd DNA, no inhibition is detected, even with a 1000-fold excess of inhibitor, when the concentration of DNA from RdSm is low (10^{-5} μg./ml.).

When high concentrations (several μg./ml.) of transforming DNA are used, however, competitive inhibition is confirmed as well with *H. influenzae*. These seemingly paradoxical results can be understood if the number of competent bacteria involved is taken into consideration. A convenient way of doing this is to express the concentration of transforming DNA in terms of input multiplicity, i.e. in number of DNA molecules present per competent bacterium. The recipient culture contains 4×10^8 viable bacteria per ml. at the peak of its competence, when all experiments are made. With a molecular weight of 15×10^6, 1 μg. of DNA corresponds to 4×10^{10} molecules. It follows that, if 100 % of the bacteria are competent, 1 μg. of DNA per ml. of culture corresponds to a multiplicity 100.

In the reaction studied, Rd × (RdSm), the number of transformants increases with DNA concentration, until the saturating concentration of 0·1 μg./ml. is reached: under the conditions of the experiment, therefore, saturation of the competent bacteria occurs at a multiplicity 10.

The results of a systematic, quantitative study of inhibition are presented in Table 4. Concentrations are expressed in 'multiplicity' for the transforming DNA, and for the inhibitory DNA in 'excess', i.e. in relative units, the concentration of transforming DNA being, in each case, taken as unity. The following observations can be made:

(1) Whatever the multiplicity of the transforming DNA, it is always possible to obtain inhibition, if sufficient inhibitory DNA is used.

(2) The competitive nature of inhibition is observed only when the multiplicity of transforming DNA is equal to, or higher than 10, i.e. when the competent bacteria are saturated with transforming DNA.

(3) When the bacteria are not saturated by transforming DNA (multiplicity < 10), the lower its multiplicity, the higher the excess of inhibitory DNA required for inhibition to take place. As a result of this, the smallest

concentration of inhibitory DNA leading to an easily detectable inhibition ($>20\%$), is *constant*, and equal to $0\cdot1$ μg./ml. or multiplicity 10. (For example, with multiplicity 10^{-3} of transforming DNA, inhibition appears when 10^4 times this concentration is given of the inhibitory DNA, i.e. for multiplicity 10, or $0\cdot1$ μg./ml., of inhibitory DNA. With multiplicity 10^{-1} of transforming DNA, inhibition appears for an excess 10^2 of inhibitory DNA, i.e. again $0\cdot1$ μg./ml. of inhibitory DNA, etc.)

Table 4. *Inhibition of transformation by varying concentrations of DNA from the recipient strain, as a function of transforming DNA concentration* (from Schaeffer, 1957b). *Transforming DNA in 'multiplicity', inhibitory DNA in 'excess', as explained in the text*

Multiplicity of transforming DNA	Percentage inhibition of transformation obtained when the excess of inhibitory DNA is equal to					
	10^0	10^1	10^2	10^3	10^4	10^5
10^{-3}	0	0	0	0*	40	82
10^{-2}	0	0	0	45	95	—
10^{-1}	0	0	35	87	—	—
10^0	0	52	93	—	—	—
10^1	42	80	98	—	—	—
10^2	31	88	—	—	—	—
6×10^2	53	—	—	—	—	—
$1\cdot8\times10^3$	49	—	—	—	—	—

* o means no clear-cut inhibition, or inhibition $< 20\%$.

In conclusion, inhibition is of a competitive nature, as can be shown when saturating concentrations of transforming DNA are used. But with concentrations of transforming DNA too low to saturate the competent bacteria, inhibition is obtained only when the competent bacteria are saturated with the inhibitory DNA, and the competitive nature of inhibition is masked. The apparent contradiction in the initial, isolated observations, vanishes when these facts are recognized.

As already mentioned, in the calculations presented, all bacteria are assumed to be competent in a fully competent culture of Rd. This, already suspected by Goodgal & Herriott (1957) on entirely different grounds, can actually be derived from the data of Table 4, as has been shown elsewhere (Schaeffer, 1957b).

The site of competition

To say that inhibition of transformation by DNA is of the competitive type is equivalent to saying that at least one specific receptor exists in a competent bacterium, the competition between DNA molecules being for combination with this receptor. The question then arises of the nature of the receptor and of its location in the bacterium. Its nature is still a matter of speculation and will be discussed later; its localization is a much simpler problem to solve.

Admitting, as is done here, that the transforming molecule must logically pair with a structure on the chromosome in order that transformation may occur, the site might be this very structure, and the competition might occur inside the bacterium. In order that an inhibitory DNA may occupy the structure and prevent transforming molecules from approaching, its pairing

Table 5. *Compared inhibitions, by non-radioactive DNA from the recipient strain, of transformation and of the retention by washed bacteria of radioactive transforming DNA* (from Schaeffer, 1957 c). *The reaction studied is* $Rd \times (^{32}P\text{-}RdSm + Rd)$

| | Percentage inhibition | |
Concentration of inhibitory DNA*	Retention, by DNase treated bacteria, of radioactive transforming DNA	Transformation
0	0	0
3	44	42
10	73	71
30	87	86

* The concentration of transforming DNA (0·1 μg./ml.) being taken as unity.

need not be good: a few points of attachment should suffice. Equally likely, however, is the possibility that, transformation being a matter of pairing and copy-error, inhibition is a matter of adsorption and penetration; in this case, the receptor would be a component of the cell wall, and the DNA molecules competing for it would be outside the bacterium.

The alternatives can be distinguished by making use of ^{32}P-labelled transforming DNA. If the competition is for adsorption, increasing concentrations of non-radioactive inhibitory DNA are expected to depress equally both the amount of radioactivity retained by the washed bacteria and the number of transformants, while the latter only is expected to be lowered if competition occurs inside the bacterium.

In the experiment, cultures of the strain Rd were given mixtures of ^{32}P-DNA from RdSm and non-radioactive DNA from Rd itself. The results, expressed in Table 5 as percentage inhibition for both retention and transformation, clearly show identical inhibitions for the two phenomena (Schaeffer, 1957 c).

Receptors, necessary for DNA adsorption and penetration, seem therefore to be present at the surface of competent bacteria. Non-transforming

DNA molecules compete with transforming ones for these receptors and thereby eventually inhibit transformation. As all the DNA's so far studied have been shown to 'penetrate' the cell equally well the receptors would seem to be devoid of species-specificity.

Discussion on the species-specificity of DNA fixation

With the information gathered so far, two predictions can be made.

Prediction 1. As we have just seen, inhibition of Rd transformation is a matter of competition for some receptor located at the surface of competent Rd bacteria. Now, DNA's from RdSm, FidSm and A61Sm have an equal affinity for these receptors (hence their equal fixation), and the same must be true of the DNA's from the corresponding wild types (see Goodgal & Herriott, 1957). It follows therefore that the DNA's from the three

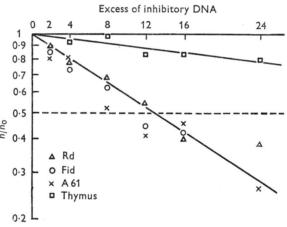

Fig. 2. Inhibition of the transformation Rd × (RdSm) by DNA's of various origin (from Schaeffer, P., *Ann. Inst. Pasteur*, in the Press). Concentration of transforming DNA: 0·01 μg./ml.; ordinates as defined in text.

wild-type strains must be equally potent inhibitors of Rd transformation, despite the very unequal ability to transform Rd exhibited by the DNA's from the three mutant strains. This prediction has been tested experimentally. Fig. 2 gives the inhibition curves obtained when transformation of Rd by DNA from RdSm (0·01 μg./ml.), is inhibited by increasing concentrations of DNA from Rd, Fid, A61 or calf thymus.

When n/n_0, the ratio of the number of transformants in the presence of inhibitor to the number in its absence, is plotted on a logarithmic scale against the excess of inhibitor, straight lines are obtained, and inhibition data with the three *Haemophilus* DNA's fall on one and the same line: the three bacterial DNA's, as expected, are equally inhibitory. The first prediction is therefore verified, the absence of correlation between the transforming and the inhibitory activities of a DNA being a reflexion of: (1) the distinctness of the sites involved in pairing and in inhibition, and (2) the difference in species-specificity of these two phenomena.

Prediction 2. Inhibition being a matter of adsorption, DNA's which are less potent inhibitors of Rd transformation than DNA from Rd itself (like DNA from calf thymus, Fig. 2), must have less affinity for the Rd receptors, and their fixation by Rd cultures must be low. As a preliminary step on the way to verification of this prediction, bacterial DNA's have been found, which are no more inhibitory of Rd transformation than thymus DNA (for instance DNA's from *Haemophilus ducreyi, H. pertussis, Brucella bronchiseptica,* etc.). This is important technically, as all bacterial DNA's can be prepared according to the same procedure, and be easily labelled. But the experimental verification of the second prediction is still lacking.

If future investigation demonstrates the correctness of the present expectation, i.e. if the observed differences in inhibitory activity of DNA's from different species, as measured on transformation of Rd, are correctly interpreted in terms of unequal affinities of these DNA's for the Rd receptor, it will have two important implications:

(1) The receptor will have to be considered as endowed with specificity. Admittedly, it does not discriminate between the DNA's from Rd (*Haemophilus influenzae*), Fid (*H. parainfluenzae*) and A61 (*H. suis*), but if it does discriminate between those and the DNA's from *H. ducreyi, H. pertussis,* etc., it would at least exhibit group specificity. This, together with the recent suggestion of Fox & Hotchkiss (1957), that the receptor might be a protein, would go a good deal of the way towards categorizing its enzymic nature.

(2) The molecular mechanism underlying the alleged pairing of an exogenous DNA molecule with a chromosomal structure is still a matter of speculation; that this pairing is extremely specific is no wonder, however, as it seems that, when it comes to pairing, the genetic information of the molecule (i.e. the succession of nitrogen bases normally hidden inside the duplex), must somehow become exposed. But the same assumption for a molecule outside the bacterium, just coming into contact with the cell-wall receptor, does not seem to be reasonable. Does that mean that there is room for specificity in the groove that runs all around the surface of a duplex molecule? Or is there a minor non-DNA component in a 'DNA' molecule, responsible for this kind of specificity? Such are the questions that study of interspecific transformation reactions is likely to lead to.

Note

In discussion Dr J. M. Thoday pointed out that it may be possible to interpret Dr Schaeffer's results without invoking different frequencies of intra- and interspecific transformation. Experience with *Drosophila* indicates that chromosomes combining parts of homologues from different populations often lead to low viability. It might be supposed that transformed bacteria may similarly be inviable if a considerable piece of foreign genetic material is transferred. Then one would only detect those of the interspecific transformations that involved small amounts of foreign genetic material. One would, of course, detect a larger proportion of intraspecific transformations, so that though intra- and interspecific transformations might be equally frequent, the observed interspecific transformations would be rarer. Those interspecific transformants that survived might well be as viable as, and behave like intraspecific transformants.

Addendum

All that can be said at present is that cell counts, made on a non-selective medium immediately after addition of DNase, are the same, whether the cultures were treated with heterospecific DNA or not. This, it is true, leaves Dr Thoday's proposed interpretation still open, even though the majority of the cells pick up DNA, as the competent cells may well contain more than one nucleus.

REFERENCES

ALEXANDER, H. E. & LEIDY, G. (1950). *Proc. Soc. Exp. Biol. Med.* **73**, 485.
ALEXANDER, H. E. & LEIDY, G. (1955). *Amer. J. Dis. Child.* **90**, 560.
ALEXANDER, H. E., LEIDY, G. & HAHN, E. (1954). *J. Exp. Med.* **99**, 505.
DEMEREC, M. & DEMEREC, Z. E. (1956). In *Mutation, 8th Brookhaven Symposium in Biology*, p. 75. Upton, New York: Brookhaven National Laboratories.
FOX, M. S. (1957). (in the Press).
FOX, M. S. & HOTCHKISS, R. D. (1957). *Nature, Lond.* **179**, 1322.
GOODGAL, S. H. & HERRIOTT, R. M. (1957). In *The Chemical Basis of Heredity*, p. 336. Ed. McElroy, W. D. & Glass, B. Baltimore: Johns Hopkins Press.
GRIFFITH, F. (1928). *J. Hyg., Camb.* **27**, 113.
HOTCHKISS, R. D. (1954). *Proc. Nat. Acad. Sci., Wash.* **40**, 49.
HOTCHKISS, R. D. (1956). In *Enzymes: Units of Biological Structure and Function*, p. 119. Ed. Gaebler, O. H. New York: Academic Press.
HOTCHKISS, R. D. (1957). In *The Chemical Basis of Heredity*, p. 321. Ed. McElroy, W. D. & Glass, B. Baltimore: Johns Hopkins Press.
LEIDY, G., HAHN, E. & ALEXANDER, H. E. (1956). *J. Exp. Med.* **104**, 305.
SCHAEFFER, P. (1956). *Ann. Inst. Pasteur*, **91**, 192.
SCHAEFFER, P. (1957a). *C.R. Acad. Sci., Paris*, **245**, 375.
SCHAEFFER, P. (1957b). *C.R. Acad. Sci., Paris*, **245**, 451.
SCHAEFFER, P. (1957c). *C.R. Acad. Sci., Paris*, **245**, 230.
SCHAEFFER, P. & RITZ, E. (1955). *C.R. Acad. Sci., Paris*, **240**, 1491.

GENETIC AND PHYSICAL DETERMINATIONS
OF CHROMOSOMAL SEGMENTS
IN *ESCHERICHIA COLI*

By FRANÇOIS JACOB and ELIE L. WOLLMAN

Service de Physiologie microbienne, Institut Pasteur, Paris

In *Escherichia coli* K-12, conjugation involves the *oriented transfer* of a chromosomal segment from a donor to a recipient bacterium. This process is oriented in the sense that the different genetic determinants located on the chromosomal segment transferred from the donor penetrate into the recipient in a predetermined order, and according to a rather precise time schedule (Wollman & Jacob, 1955). This characteristic of the mating process makes conjugation in bacteria a suitable material for relating genetic analysis to genetic structures, and for comparing genetic evaluation of these structures with physical measurements.

After summarizing our present knowledge of the process of conjugation in *E. coli* K-12, it is intended, in the first part of this paper, to report the available information on the organization of the genetic material in this organism. The second part will be mainly concerned with the effects of ultra-violet light and of disintegration of radioactive phosphorus on the processes of conjugation and recombination. These last experiments allow, as will be discussed, a comparison between genetic and physical determinations of chromosomal segments in bacterial crosses.

I. THE GENETIC SYSTEM OF *E. COLI* K-12

The process of conjugation

When two strains of *E. coli* K-12, which differ in such properties as synthesis of essential metabolites, fermentation of sugars, resistance or sensitivity to bacteriophages or to drugs, are mixed, genetic recombination may be demonstrated between characters of the parental types. In the recombinants thus formed the characters of the parents are reassorted and certain types of recombinants may be easily scored by plating on suitable selective media (Tatum & Lederberg, 1947; Lederberg, 1947). Sexual differentiation in *E. coli* K-12 has been demonstrated, and genetic recombination involves the transfer of genetic material from *donor* to *recipient* bacteria (Hayes, 1953a; Cavalli, Lederberg & Lederberg, 1953). Whereas no essential difference has been recognized between different strains of recipient (or, F^-) bacteria, two main types of donors may be distinguished. Most of the K-12 strains, including the wild type, are F^+: upon mixing with F^- cells they exhibit a low frequency of recombination (10^{-5} or less of any type of recombinant). Some strains, however, exhibit a high frequency of recombination (from 10^{-1} to 10^{-3}) and are thus called *Hfr* (Cavalli, 1950; Hayes, 1953b). Closer analysis shows that only certain characters of an *Hfr* strain are transmitted at high frequency to recombinants, whereas others are transmitted at the same low frequency as is observed in $F^+ \times F^-$ crosses (Hayes, 1953b).

The high frequency of recombination that may be observed in $Hfr \times F^-$ crosses allows an analysis of the process of conjugation at the cell level (Wollman, Jacob & Hayes, 1956). One may schematically distinguish several successive steps in this process (Jacob & Wollman, 1955).

(i) The first step consists in the establishment of an *effective contact* between cells of opposite mating types. Under suitable conditions of environment and cell density this step is rapidly completed and practically all possible matings occur within 30 min. after mixing of bacteria of opposite mating types. Electron micrographs demonstrate the existence of a bridge which unites conjugating bacteria (Anderson, Wollman & Jacob, 1957).

(ii) The second step consists in the oriented *transfer* of a chromosomal segment of the *Hfr* donor into the F^- recipient. The mechanism of transfer has been analysed by interrupting the process of conjugation at different times after its onset, by means of a Waring blendor (Wollman & Jacob, 1955). It was thus found that the genetic characters linearly arranged on the *Hfr* chromosome segment penetrate into the recipient in a predetermined order and always the same extremity, O (for origin), first. This process is slow enough to be interrupted at various times by mechanical treatment. Interruption of the mating process does not prevent any genetic character which has already penetrated the recipient from being later integrated into a recombinant. Even when the process is not artificially interrupted, spontaneous breaks do occur during transfer of the *Hfr* chromosome, with the result that the length of the segment transferred varies from one mating pair to the other. As a result of transfer a partial zygote or *merozygote* is formed, which comprises the whole F^- recipient and a chromosomal segment of *Hfr* origin.

(iii) The third step involves genetic recombination proper, that is the series of events which lead to the *integration* of *Hfr* markers to form a recombinant chromosome.

(iv) The fourth step, or *expression*, comprises the events which, through segregation and phenotypic expression, extend from the formation of a recombinant chromosome to that of a fully expressed recombinant bacterium.

Of these four main steps, the last two, integration and expression, are common to the different known processes of genetic transfer in bacteria, i.e. conjugation, transformation and transduction. Analysis of the results obtained in the study of bacterial conjugation involves the comparison, in each case, of the nature and extent of genetic transfer from donor bacteria to the zygotes with the subsequent integration of these characters to recombinants. The characters transmitted with high frequency by strain *HfrH*, the strain originally described by Hayes (1953*b*) are represented in Fig. 1. The order and relative distances of the characters located on this segment may be determined, as indicated in the legend, both by the genetic analysis of recombinants and by the time at which these characters penetrate, during transfer, into the F^- recipient.

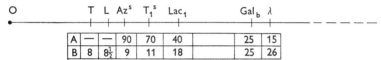

Fig. 1. Genetic map of the segment injected with high frequency by the *Hfr* isolated by Hayes. The location of the different characters as measured in crosses *Hfr* × *F*⁻ by *A*: the percentage of T⁺L⁺Sʳ recombinants which have inherited the different *Hfr* alleles; *B*: the time at which individual *Hfr* characters start penetrating into the *F*⁻ recipient in a Waring blendor experiment.

The mating systems of E. coli *K-12*

When an attempt is made to compare conjugation between *Hfr* and F^- bacteria on the one hand, and between F^+ and F^- bacteria on the other, the question arises as to whether the low frequency of recombination observed in the latter crosses is a consequence of a low frequency of conjugation, of transfer, or of integration. In $F^+ \times F^-$ crosses, the F^+ character itself is transmitted at high frequency to F^- cells (Cavalli *et al.* 1953) and this is also the case for the ability to produce certain colicins (Fredericq & Betz-Bareau, 1953). This indicates that the frequency of effective contacts, and hence of conjugation, is as high in $F^+ \times F^-$ crosses as in $Hfr \times F^-$ crosses (Jacob & Wollman, 1955), a prediction which is verified by microscopic studies (Anderson *et al.* 1957). Other lines of evidence support the hypothesis that low frequency of recombination is a consequence of a low frequency of transfer.

The question therefore arises of the origin of those recombinants which are formed at low frequency in $F^+ \times F^-$ crosses. They could either result from a low but constant probability for each F^+ cell to transfer the character under consideration—or from the presence, in F^+ cultures, of a small proportion of *Hfr* mutants of variable nature. Evidence for the validity of the latter hypothesis comes from both the quantitative and qualitative results of 'fluctuation tests' as well as from the possibility of isolating, with a high yield, those *Hfr* mutants which are responsible for the formation of recombinants of any type in $F^+ \times F^-$ crosses (Jacob & Wollman, 1956*a*).

It may be pointed out that both mechanisms could play a part in the formation of recombinants in $F^+ \times F^-$ crosses. The demonstration that most of these recombinants, if not all of them, are formed by *Hfr* mutants present in the F^+ population, however, indicates that this mechanism is the prevailing one. If any recombinant were formed by F^+ donors, the detection of this latter mechanism would be extremely difficult, and there is, at the present time, no experimental evidence for its existence.

The patterns of chromosome transfer in Hfr mutants

The finding that most, if not all, recombinants formed in an $F^+ \times F^-$ cross are due to *Hfr* mutants has two main implications. On the one hand, the fact that any known marker of an F^+ may be transmitted to a recombinant, implies that there must exist *Hfr* mutants able to transfer such markers at high frequency. On the other hand, the fact that any *Hfr* strain, such as the strain of Hayes, may transfer at high frequency only a group of characters suggests that the different *Hfr* mutants present in an F^+ culture must differ as to the nature of the chromosomal segment they are able to transfer.

This is indeed what is found. Any known genetic character of *E. coli* K-12 may be transferred at high frequency by a given type of *Hfr* mutant. These mutants differ from each other in the nature of the chromosomal segment they are able to transfer at high frequency. No *Hfr* mutant has been isolated so far which can transfer at high frequency all the genetic markers known (Jacob & Wollman, 1956*a*).

When studying any *Hfr* strain, one may obtain, as shown in previous sections, two types of information. On the one hand, one may determine, by genetic analysis, which characters are transmitted at high frequency to recombinants as well as their relative frequency of transmission. On the other hand, one may determine, by a blendor experiment, the sequence in which these characters are transferred and the time at which any given character enters the recipient.

Such an analysis has been carried out with a variety of *Hfr* mutants isolated from different strains of *E. coli* K-12 (Jacob & Wollman, 1957). Preliminary results are summarized in Table 1. It is seen that different *Hfr* strains differ from each other not only in those *characters* that they are able to transfer at high frequency, but also in the *order* in which any group of characters may be transferred. The simplest representation of these results consists in assuming that, for any given *Hfr* strain, a chromosomal segment is transferred through a given extremity O (the origin) and that it is the position of O which determines which characters are transferred at high frequency, as well as the order of their transfer. A remarkable feature of the genetic system of *E. coli* K-12 appears to be the existence of a predetermined pattern of arrangement of the genetic characters that an *Hfr* mutation will affect by determining the position of O and hence the orientation of transfer. When considering a given genetic character B, it is found to be linked to a character A on one side and to a character C on the other side, these linkage relationships being retained as long as O is not

Table 1. *Patterns of Hfr mutants*

Schematic representation of the characters injected with high frequency by some of the isolated *Hfr* mutants. Each line corresponds to an *Hfr* strain and the order of injection corresponds to the characters from left to right.

Type	O←—————————————————————————————————
H	TL Az T_1 Lac T_6 Gal λ
1	L T B_1 M Mtol Xyl Mal S^r
2	T_1 Az L T B_1 M Mtol Xyl Mal S^r
3	T_6 Lac T_1 Az L T B_1 M Mtol Xyl Mal S^r
4	B_1 M Mtol Xyl Mal S^r λ Gal
5	M B_1 TL Az T_1 Lac T_6 Gal λ

located closer to B than either A or C. When B becomes linked to O on one side, either it remains linked to A, and then becomes completely unlinked to C, or it remains linked to C and then becomes completely unlinked to A. For instance, in an *Hfr* of type 5 (Table 1), the characters TL appear to be linked to M (methionine) on the one side and to Gal on the other side, the order of transfer being O-M-TL-Gal. In the classical *HfrH*, characters TL are now linked to O on one side and to Gal on the other side, the order of transfer being O-TL-Gal. On the contrary, in *Hfr* of type 2, TL are linked on one side to O, but on the other side they are linked to M and the sequence of transfer has now become O-L-T-M.

Such a relationship is valid for all the known characters of K-12 and for all the *Hfr* strains which have been isolated up to now, whatever the previous history of the F^+ strains of K-12 from which they originate.

When trying to draw a genetic map of K-12, one is thus faced with a paradoxical situation. On the one hand, we do not possess any direct information on the genetic system of either F^+ or F^- bacteria. On the other hand, when analysing any given $Hfr \times F^-$ cross, we can only determine a genetic map of the chromosomal segment transferred at high frequency by that *Hfr* strain. We are therefore compelled to compare the genetic segments thus determined. From such a comparison, the conclusion is reached that all the known genetic characters of *E. coli* K-12 are linked, a conclusion which was already attained by the analysis of $F^+ \times F^-$ crosses (Clowes & Rowley, 1954; Cavalli & Jinks, 1956), but that they cannot be arranged along a straight line as is the case in most chromosomes. There is no reason in effect for interrupting the determined linkage group at any place preferentially to any other one, except when describing the properties of a particular *Hfr* strain. One is thus led to dispose all the

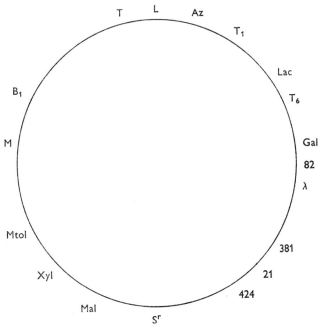

Fig. 2. Diagrammatic representation of the K-12 chromosome, as it results from a comparative study of the segments injected with high frequency by different *Hfr* strains. This diagram represents only the sequence of characters, not the distances between them. Symbols refer to threonine (T), leucine (L), methionine (M) and thiamine (B_1) synthesis; resistance to sodium azide (Az), bacteriophages T_1 and T_6, streptomycin (S); fermentation of lactose (Lac), galactose (Gal), maltose (Mal), xylose (Xyl) and mannitol (Mtol); ultra-violet inducible prophages 82, λ, 381, 21 and 424. (Possible mechanism of the $F^+ \rightarrow Hfr$ mutation: insertion of a specific factor at a given place of the circular linkage-group would result in the interruption of the circle. One extremity of the linkage group would behave as the origin. The other would carry the *Hfr* character.)

known genetic characters on a circle (Fig. 2). Since all *Hfr* mutants derive from F^+ strains, the underlying pattern of all *Hfr* would be that of the F^+ genetic system. It seems unnecessary to emphasize that this diagrammatic representation, which is the simplest one that will account for the observed results at the present time, is not meant to imply that the bacterial chromosome is actually circular.

Thus we have, on the one hand, in every individual *Hfr* a linear representation, one extremity of which is perfectly determined by the origin O, and, on the other hand, a circular representation for the whole sequence. Under this scheme, the pattern of any existing *Hfr* may be obtained by interruption of the circle at the proper place, one of the extremities thus formed becoming the origin O.

Once the complete sequence of the K-12 linkage group has been determined by comparing the segments transmitted with high frequency by the various *Hfr* strains, one may analyse the capacity of a given *Hfr* type to transmit to recombinants those characters which appear only with low frequency. It seems that any *Hfr* type is indeed able to transmit all the known characters, but the frequency with which a given character is transmitted decreases the further from O this character is located in the particular *Hfr* strain. Such a genetic polarity may equally be interpreted as resulting from a gradient of pairing (Cavalli & Jinks, 1956) or from a gradient of transfer (Wollman & Jacob, 1957). The available evidence seems to indicate that the genetic polarity does reflect a polarity of transfer. The further from O a character is located, the greater the chance that a break will prevent its transfer during conjugation. When characters are located at a given distance from O, the probability for their transfer becomes so small that they appear very rarely among recombinants. These are the characters which are said to be transmitted to recombinants at low frequency in an *Hfr* × *F⁻* cross.

Among characters transmitted with low frequency seems to be the *Hfr* character itself. Several of the isolated *Hfr* strains have been analysed for their capacity to transmit their *Hfr* character. With these strains, the recombinants having inherited from the *Hfr* parent the characters located close to O and transmitted with high frequency (proximal characters) are all *F⁻*. On the contrary, among the few recombinants which have inherited from the *Hfr* parent characters located close to the extremity opposite to O (terminal characters), many are *Hfr*. When crossed with *F⁻* bacteria, these *Hfr* recombinants appear to inject their markers in the same order as the one found for the *Hfr* parent from which they were derived. These properties of *Hfr* strains are similar to those already described for the two *Hfr* strains which were first isolated (Hayes, 1953*b*; Cavalli & Jinks, 1956). It appears, therefore, that the *Hfr* character does segregate among recombinants but that its linkage to other markers depends upon the strain considered. In any given *Hfr* strain, the *Hfr* character seems to be located at, or close to, the terminal extremity of the linkage group.

If confirmed by further investigation, this could be interpreted by the assumption that the event, which, in the *F⁺*→*Hfr* mutation, is supposed to result in the rupture of the circle at a given point, not only would determine the position of O at one of the two ends of the linkage group, but also the location of the *Hfr* character at the other end. The properties of the different *Hfr* strains could thus be accounted for by the single hypothesis that the insertion of a specific factor at the proper place in the circular linkage group would determine the rupture of the circle. One extremity of the linkage group would behave as the origin O and would be injected during the process of mating. The other extremity would be terminal and would carry the *Hfr* character.

Again it must be clearly stated that this hypothesis provides only a formal model and that other schemes might well explain the experimental results. The proposed model, however, appears to be the simplest one that accounts for all the data. Although the nature of the postulated factor, whose insertion would be responsible for the breaking of the circle, remains unknown, it might well be conceived as being similar to the 'controlling elements' described in maize by McClintock (1956).

One must conclude that any *Hfr* may, upon conjugation, inject into the F^- recipient, a particular sequence of genetic characters in an oriented way. The distance between these characters may be determined by genetic analysis, when these characters are not very far apart. For distant characters, genetic analysis is grossly deformed by the increasing probability of breaks occurring, during transfer, between these characters. A more precise determination of the distances may be obtained by measurement of the time at which different characters penetrate the F^- recipient during conjugation.

II. EFFECT OF ULTRA-VIOLET LIGHT AND ^{32}P DECAY ON BACTERIAL RECOMBINATION

Radiation and disintegration of radioactive phosphorus are known to induce various types of cellular lesions and more particularly local alterations in the genetic material. In bacterial conjugation, whereas the F^- recipient contributes to the zygote both its genetic material and its cytoplasm, the *Hfr* donor appears to contribute genetic material but very little, if any, cytoplasm. Exposure of the *Hfr* donor to physical agents is therefore likely to affect only one of the genetic elements taking part in recombination. Lesions induced in the *Hfr* may affect either the transfer of genetic material or its integration in the zygote after transfer.

As shown in previous sections, only a segment of the *Hfr* chromosome may be investigated in a cross between a given *Hfr* strain and a recipient F^-. The following discussion will only be concerned with the segment injected at high frequency by the strain *HfrH* isolated by Hayes (Fig. 1).

When crosses are performed between *HfrH* $T^+L^+Az^sT_1^sLac_1^+Gal_b^+S^s$

and *P678* $F^-T^-L^-Az^rT_1^r$ Lac_1^- Gal_b^- S^r,* a chromosomal segment O-TL-Gal (Fig. 1) of the *Hfr* is transmitted at high frequency. The S^s marker not being transferred at high frequency, streptomycin may be used for eliminating the *Hfr* parent after mating. By plating the zygotes on suitable selective media, one may select, in the same experiment, recombinants which have inherited certain characters of the *Hfr* parent, either *proximal* to O, such as T^+L^+ which are close enough to be used together as selective markers (recombinants $T^+L^+S^r$) or distal to O such as Gal^+ (recombinants Gal^+S^r). One may also select those recombinants which have received the T^+L^+ as well as the Gal^+ characters of the *Hfr* (recombinants $T^+L^+Gal^+S^r$). In such an experiment, one may compare the frequencies with which these different types of recombinants are formed as well as their genetic constitution, that is the distribution among these recombinants of the genetic markers contributed by the *Hfr* parent.

The effect of ultra-violet irradiation

If *Hfr* donors are first exposed to ultra-violet light and then mated with non-irradiated F^- recipients, the irradiated *Hfr* lose their ability to form recombinants exponentially as a function of the dose of ultra-violet (Fig. 3). The capacity to form either $T^+L^+S^r$ recombinants or Gal^+S^r recombinants is lost at the same rate. This indicates that the transfer of the distal markers is not more reduced than that of the proximal ones. It is not known whether ultra-violet light may or may not prevent transfer, but if it does, it is not by reducing the size of the transferred piece.

On the contrary, the capacity to form recombinants which have received *both* extremities of the TL-Gal segment ($T^+L^+Gal^+S^r$ recombinants) is lost at a faster rate. One must, therefore, conclude that lesions produced in the *Hfr* parent before mating alter the process of integration occurring in the zygote after transfer. They decrease the probability of simultaneous integration of distant markers.

The difference observed between the slopes of the $T^+L^+S^r$ and $T^+L^+Gal^+S^r$ recombinants indicates that the fraction of the $T^+L^+S^r$ recombinants which have inherited the Gal^+ marker from the *Hfr* parent decreases as a function of the dose of ultra-violet. One may therefore expect to observe an alteration of the genetic constitution of the recombinants formed by irradiated *Hfr* parents. As shown in Table 2, the frequency with which unselected markers of the *Hfr* are present in both $T^+L^+S^r$ and Gal^+S^r recombinants is strikingly reduced by irradiation. This means that irradiation of the *Hfr* parent results in a loosening of the linkage observed

---

* Synthesis of threonine (T), leucine (L), fermentation of lactose (Lac), galactose (Gal), sensitivity (s) or resistance (r) to sodium azide (Az), to phage T1 and to streptomycin (S).

between the markers of the TL-Gal segment. In other words, the probability of a crossover occurring between two markers is increased by irradiating the donor.

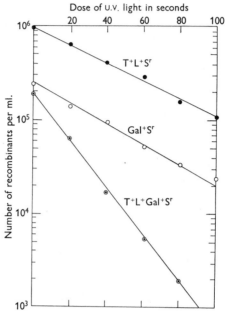

Fig. 3. Recombination between ultra-violet irradiated *Hfr* donor and non-irradiated *F⁻* recipient. A suspension of *HfrH* is exposed to various doses of ultra-violet light. Samples of irradiated suspensions are mixed in standard conditions with non-irradiated *P678 F⁻* recipients and aliquots are plated on different selective media. The number of recombinants T⁺L⁺Sʳ, Gal⁺Sʳ and T⁺L⁺Gal⁺Sʳ/ml. of mating mixture are plotted on a logarithmic scale versus the dose of ultra-violet light in seconds.

Table 2. *Effect of ultra-violet irradiation on recombination*

Genetic composition of the recombinants formed in different crosses between *HfrH* and *P678 F⁻*. In the two last crosses, one of the parents was submitted to a standard dose of ultra-violet leaving about 30–50 % survivors. The figures represent the percentage of recombinants T⁺L⁺Sʳ or Gal⁺Sʳ having inherited the characters Az, T_1, Lac and Gal of the *Hfr* parent.

| | Genetic constitution of recombinants | | | | | | | |
| | T⁺L⁺Sʳ | | | | Gal⁺Sʳ | | | |
Crosses	Az	T_1	Lac	Gal	TL	Az	T_1	Lac
Hfr × F⁻ control	91	72	52	29	74	78	73	74
Hfr u.v. × *F⁻*	52	34	11	3	16	21	29	36
Hfr × F⁻ u.v.	88	60	35	16	41	49	57	62

Irradiation of the recipient also exerts some effect on the genetic constitution of recombinants, but to a much smaller extent than irradiation of the donor (Table 2).

This action of ultra-violet light on bacterial recombination appears to be similar to the effect which the same doses of ultra-violet exert on phage recombination (Jacob & Wollman, 1955). Both effects can be best interpreted according to the hypothesis that, in bacteria as well as in phages, genetic recombination does not occur by breakage and reunion of complete strands but rather by a mechanism connected with replication, which, having commenced on one chromosome, would shift and be finished on the other (Levinthal, 1954).

Lesions produced randomly on the irradiated genetic material would interfere with the process of replication and therefore increase the frequency of the replication shifting from one parental chromosome to the other. This would result in a decrease of the size of the piece, or pieces, contributed by the irradiated parent to recombinants. Hence an apparent stretching of the linkage group.

The effect of ^{32}P disintegration

It is known that, like bacteriophages (Hershey, Kamen, Kennedy & Gest, 1951), bacteria containing ^{32}P of high specific radioactivity lose their viability as a function of ^{32}P decay (Fuerst & Stent, 1956). This method is of special interest since the lethal effects of ^{32}P decay cannot be accounted for by the ionizations produced, but appear rather to result mainly from 'short range' consequence of radioactive disintegrations, such as the transmutation ^{32}P$\rightarrow^{32}$S or the recoil energy sustained by the decaying P nucleus (Hershey *et al.* 1951; Stent, 1953). The favoured hypothesis is that the lethal effect of ^{32}P decay is due to those disintegrations which produce a rupture in the DNA chain.

This method may be applied to the study of bacterial recombination by mating ^{32}P-labelled *Hfr* donors with non-radioactive *F⁻* recipients. By comparing the results of two types of experiments, it is possible to distinguish between the effects of ^{32}P decay on integration and its effects on transfer.

In one type of experiment radioactive *Hfr* donors are mated with non-radioactive recipients and ^{32}P decay allowed to occur *in the zygotes just after mating*, by storage in the cold. The capacity of the zygotes to give rise to recombinants is then measured as a function of ^{32}P decay. This type of experiment makes it possible to measure the effect of ^{32}P decay on *integration* independently of any effect on transfer.

The second type of experiment consists in allowing ^{32}P decay to occur *in the Hfr donors before mating*. Radioactive *Hfr* bacteria are stored in the cold and their *mating capacity*, that is their ability to transmit genetic characters to recombinants upon mating with *F⁻* non-radioactive recipients, is measured as a function of ^{32}P decay. In this type of experiment radioactive disintegration may affect both the transfer of the genetic material and its integration (Fuerst, Jacob & Wollman, 1956, 1958).

(1) *The effects of* ^{32}P *decay occurring in the zygotes after mating*. When radioactive *Hfr* donors are mated with non-radioactive F^- recipients in non-radioactive medium and the early zygotes formed are stored in the cold, it is found that the capacity of these zygotes to form recombinants of any type decreases as a function of the time of storage, that is of ^{32}P decay (Fig. 4). This instability of the capacity of the early zygotes formed indicates that the genetic segment contributed by the *Hfr* parent remains susceptible to radioactive disintegration even after its transfer to the zygote. This is proof that the effects of ^{32}P decay are indeed a consequence of events occurring in the genetic material.

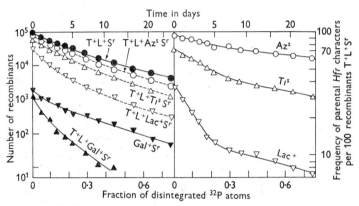

Fig. 4. Effect of ^{32}P decay occurring in the zygotes. Bacteria *HfrH* T$_6$S^s grown in a medium containing 110 mc./mg. of radio-phosphorus are washed, resuspended in buffer and mixed with an excess of non-radioactive *P678* F^- T$_6^r$S^r, in a medium containing streptomycin, to prevent the *Hfr* parent from synthesizing nucleic acids. After 40 min. at 37°, KCN M/100 and an excess of phage T$_6$ are added to stop conjugation (Hayes, 1957). After 10 min. at 37°, the mixture is diluted in protective medium and samples are frozen in liquid nitrogen. Every day a sample is thawed and aliquots are plated on selective media.

The number of recombinants T$^+$L$^+$S^r, Gal$^+$S^r and T$^+$L$^+$Gal$^+$S^r (left, solid lines) and the *proportion* of T$^+$L$^+$S^r recombinants having one of the *Hfr* characters Az^s, T$_1^s$, Lac$^+$ and Gal$^+$ (right) are plotted on a logarithmic scale versus the time in days and the fraction of disintegrated ^{32}P atoms. On the left figure are also plotted in dotted lines the *numbers* of recombinants T$^+$L$^+$S^r having one of the *Hfr* characters Az^s, T$_1^s$, Lac$^+$ and Gal$^+$, as calculated from the curves on the right.

When, in the same experiment, zygotes are sampled at different time intervals after mating, it is found that the capacity of the zygotes to form recombinants is the less sensitive to ^{32}P decay the later the time of sampling and that it becomes practically insensitive for the zygotes sampled after 120 min. This indicates that the genetic information carried by the injected chromosomal segment of the radioactive *Hfr* has been transferred to non-radioactive material.

Genetic analysis of the recombinants formed by early zygotes in the course of ^{32}P decay gives the following information. First of all, as may be seen on Fig. 4, the number of recombinants which inherit either the T^+L^+ characters or the Gal^+ character of the *Hfr* parent decreases at about the same rate, whereas the number of those recombinants which inherit *both* the T^+L^+ and the Gal^+ characters ($T^+L^+Gal^+S^r$ recombinants) decreases at a much faster rate. These results, which are comparable to those obtained after ultra-violet treatment, indicate that the effect of ^{32}P decay in the zygote is to decrease the integration of the transferred genetic characters. Analysis of the $T^+L^+S^r$ recombinants shows that the linkage of the T^+L^+ characters to any of the unselected markers located on the TL-Gal segment of the *Hfr* donor such as Az^s, T_1^s, Lac^+ or Gal^+ is the more sensitive to ^{32}P decay the farther the marker is from T^+L^+. The initial slopes of the curves thus obtained depend on the order and relative distances of the characters located on the TL-Gal segment. These results suggest that ^{32}P disintegration destroys the integrity of genetic segments and that the sensitivity of any genetic segment is roughly proportional to its length.

It may be seen in Fig. 4 that the sensitivities of the linkages to T^+L^+ of the different markers decrease as a function of the time of storage and finally tend to a common slope. The explanation for this fact appears to be that the recombinants scored after a long time of storage correspond to multiple crossovers which simulate a reduction in size of the contribution of the *Hfr* chromosome.

(2) *The effects of* ^{32}P *decay occurring in the* Hfr *before mating.* When radioactive *Hfr* donors are stored and, after various time intervals, are mated with non-radioactive F^- recipients in non-radioactive medium, it is found that the capacity of these *Hfr* to form recombinants of any type decreases exponentially as a function of the time of storage (Fig. 5). However, their capacity for transmitting to recombinants the characters T^+L^+, *proximal* to O, decreases at a rate which is about one-third of the rate at which their capacity to transmit the *distal* Gal^+ character decreases. Moreover the capacity for transmitting *both* the T^+L^+ and the Gal^+ characters ($T^+L^+Gal^+S^r$ recombinants) is lost at the same rate as the capacity to transmit the Gal^+ character alone (Gal^+S^r recombinants). This suggests that when ^{32}P disintegration takes place in the *Hfr*, it affects not only the process of integration but also the transfer of genetic material from the donor to the recipient. Everything happens as though breaks had occurred on the *Hfr* genetic segment, certain *Hfr* cells still being able to transfer a proximal piece but not a distal one. Such an interpretation of the results is in agreement with the hypothesis according to which the lethal effects of ^{32}P decay result from interruptions on the DNA chain (Stent & Fuerst, 1955).

Analysis of the T⁺L⁺Sʳ recombinants as to the presence of the genetic markers of the TL-Gal segment of the *Hfr* reveals that linkages T⁺L⁺Azˢ, T⁺L⁺T₁ˢ, T⁺L⁺Lac⁺, etc. are the more sensitive, the farther away the marker is, considered from the T⁺L⁺ selected characters (Fig. 5). By comparing the rates of survival to ³²P decay of the linkages considered, one may determine the order and relative distances of the markers. The fact, however, that the radioactive *Hfr* lose their capacity for transmitting both the T⁺L⁺ and Gal⁺ characters together (T⁺L⁺Gal⁺Sʳ recombinants) at the same rate as they lose their capacity of transmitting the Gal⁺ character alone (Gal⁺Sʳ

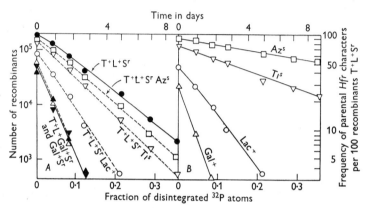

Fig. 5. Effect of ³²P decay occurring in the *Hfr* before mating. Bacteria *HfrH* grown in a medium containing 70 mc./mg. of phosphorus are centrifuged, washed and resuspended in protective medium. Samples are frozen in liquid nitrogen. Every day a sample is thawed, resuspended in buffer and mixed in standard conditions with an excess of non-radioactive recipient *P678* F⁻ (2·10⁶ *Hfr* and 10⁸ F⁻/ml.). After one hour at 37°, samples are diluted and aliquots are plated on selective media.

The number of recombinants T⁺L⁺Sʳ, Gal⁺Sʳ and T⁺L⁺Gal⁺Sʳ (left solid lines) and the *proportion* of T⁺L⁺Sʳ recombinants having one of the *Hfr* characters Azˢ, T₁ˢ, Lac⁺ (right) are plotted on a logarithmic scale versus the time in days and the fraction of disintegrated ³²P atoms. On the left figure are also plotted in dotted lines the *numbers* of T⁺L⁺Sʳ recombinants having one of the *Hfr* characters Azˢ, T₁ˢ, Lac⁺ as calculated from the curves on the right.

recombinants), indicates that the slopes of the curves of Fig. 5 represent the rate of survival of the capacity of the *Hfr* to transmit any given character to a recombinant. Since these survival curves remain exponential over a long period, the mechanism of this loss in transmissibility of genetic characters appears to be mainly, if not exclusively, an exponential decrease of their transfer. The slopes of these survival curves would then be proportional to the distance of the various characters from the extremity O of the chromosomal segment. When the relative rates of decay of the individual markers are compared (T⁺L⁺ 0·35, Azˢ 0·39, T₁ˢ 0·43, Lac⁺ 0·67, Gal⁺ 1), they are found indeed to be very similar to the relative times of entry of the same markers in a blendor experiment (T⁺L⁺ 0·34, Azˢ 0·36, T₁ˢ 0·44, Lac⁺ 0·72, Gal⁺ 1).

Disintegration of radiophosphorus therefore allows the determination of the distances between genetic markers and the genetic maps which may thus be drawn of chromosomal segments transferred at high frequency by *Hfr* strains, are in every respect comparable with the maps, which may be determined on the one hand by genetic analysis and on the other hand by the time at which genetic characters are transferred in the course of conjugation.

III. DISCUSSION

Recombination of genetic characters in bacteria may be achieved by different processes which all involve the transfer of genetic material from a donor into a recipient bacterium. In the three known processes in which this genetic transfer occurs there is little doubt that the information is carried by DNA. In transformation, pieces of naked DNA are directly taken in from the culture medium into the recipient bacteria. In transduction, it is apparently a piece of bacterial DNA which has become included within the protein coat of a phage and which, together with the DNA of the phage, is injected into the recipient bacterium upon infection. In bacterial conjugation, the process of genetic transfer is accomplished by a more elaborate mechanism, which involves contact between bacteria of opposite mating types and the oriented injection of a chromosomal segment from the donor to the recipient. Non-chromosomal material, such as an enzyme or vegetative phage, has been found not to be transferred to any measurable extent during conjugation. Although such evidence is not definite, it suggests that, if there is any transfer of cytoplasmic constituents, it must be limited as compared with that of the genetic material. That this genetic material is nucleic acid is indicated by the effects on the formation of recombinants of ^{32}P disintegration occurring in the zygotes when ^{32}P-labelled *Hfr* is mated with non-radioactive recipients.

The most characteristic feature of bacterial conjugation is the progressive transfer of an oriented chromosomal segment of the donor which always injects the same extremity first into the recipient. The polarity of the transfer offers an opportunity of determining the position and distances of the genetic characters located on the chromosomal segment transferred by different independent methods. We will briefly compare the results obtained therewith.

Genetic analysis would allow an evaluation of the distances between characters in terms of frequency of recombination between such characters. However, the frequencies thus determined may vary according to the class of recombinants selected. This results both from the incomplete nature of the zygotes which contain the complete genome of the recipient and only a fragment of the donor's, and from the variability in length of this fragment among different zygotes. By selecting recombinants for characters proximal to O (such as the $T^+L^+S^r$ recombinants obtained with *HfrH*), the recombination frequencies which may be measured (Fig. 1) would reflect both the real probability of recombination occurring between two characters and the heterogeneity in size of the segments transferred. Correction for this latter effect, which is formally comparable to a defect in pairing of homologous chromosomes (Cavalli & Jinks, 1956), may be obtained by different methods. One of them is to select those zygotes which have received chromosomal segments of equal length. This can be achieved by comparing normal crosses with crosses between lysogenic *Hfr* and non-lysogenic F^- where zygotic induction of the prophage selects those zygotes which have not received the prophage (Jacob & Wollman, 1956*b*; Wollman & Jacob, 1957). Another method consists in assuming a constant probability of breakage per length unit of the chromosome and determining the correction factor by trial and error. By combination of these two methods corrected values of the distances between genetic characters have been obtained which seem in good agreement with the results obtained with other methods (Jacob & Wollman, unpublished).

Blendor experiments permit a representation of the distances between genetic characters in time units by comparing their relative time of penetration into the recipient. The validity of this representation depends upon whether or not the transferred segment proceeds at a constant rate. Comparison between experiments done at different temperatures (Fisher, 1957), and even more the comparison between blendor and ^{32}P disintegration experiments indicate that this is the case for the characters located on the TL-Gal segment, that is along a segment which represents about one-third of the total chromosome length. The rate of penetration appears to decrease for characters located farther. It may be calculated that penetration of the whole chromosome at a constant rate would take about a hundred minutes.

Disintegration of radiophosphorus ^{32}P in the *Hfr* prior to mating appears to prevent the genetic markers from being transferred at a rate proportional to their distance from the origin O, that is to the number of phosphorus atoms existing between O and the marker considered. The genetic lesions provoked by ^{32}P decay appear, therefore, to be the direct consequence of the $^{32}P \rightarrow {}^{32}S$ transmutations which occur in the phosphodiester bonds which form the backbone of the nucleotide chain of DNA. On assuming that the efficiency of any radioactive disintegration is identical all along the chromosomal segment, a direct relationship may be established between the genetic and the physical structures of this segment. However, in order to evaluate the distances between markers in terms of number of phosphorus atoms, one must know what is the efficiency with which ^{32}P disintegration prevents transfer. In the absence of any direct information on this efficiency, the assumption may be made, that it is the same as for the lethal effects in bacteria and in phage.

It becomes thus possible to compare the different measures of the chromosome of *E. coli* K-12. It is thus found that one minute in penetration corresponds roughly to about 20 corrected recombination units and to about 10^5 nucleotide pairs. This last figure is in agreement with an independent estimate of the DNA content of an *E. coli* nucleus, which would amount to about 10^7 nucleotide pairs, and of the time of transfer of the whole chromosome at a constant rate, which would take about 100 min. The three independent methods of measuring genetic segments in *E. coli* K-12 are therefore in reasonable agreement.

IV. SUMMARY

(1) Conjugation between *Hfr* donor and *F⁻* recipient corresponds to the oriented transfer, under a rather precise time schedule, of a chromosomal segment from the donor to the recipient. The order and the distance between the characters located on the segment may be determined both by genetic analysis and by the time at which individual *Hfr* characters penetrate into the recipient.

(2) Various types of *Hfr* mutants may be isolated which differ by the characters they are able to transfer with high frequency and by the order in which the characters are transferred.

(3) By comparing the segments transferred with high frequency by various *Hfr* mutants, it is possible to determine a complete sequence of the known markers. Since no interruption can be observed, no extremity can be determined in the linkage group. One is thus led to dispose all the known genetic characters on a circle. In this model *Hfr* mutants would result from the insertion of a factor and the consequent interruption of the circle.

(4) Exposure of the *Hfr* donors to ultra-violet light before mating impairs integration of the characters located on the transferred segment. The number of genetic exchanges occurring between two given characters is increased by irradiation. This suggests that recombination in bacteria does not occur by breakage and reunion but rather by some 'copying choice' mechanism.

(5) If *Hfr* donors containing ^{32}P are mated with non-radioactive recipients, and if the mating process is interrupted early in order to allow

^{32}P decay to occur, the zygotes lose, as a function of time, their capacity to produce recombinants. Radiophosphorus disintegration appears to act by altering integration of the transferred *Hfr* markers.

(6) If radioactive *Hfr* are stored, in order to allow ^{32}P decay to occur before they are mated with non-radioactive recipients, they lose, as a function of time, their capacity to produce recombinants. Radiophosphorus disintegration appears to act by altering both transfer and integration of genetic material.

(7) It is possible to determine, by this method, the order and the distances between the markers located on the *Hfr* segment transferred with high frequency. The genetic map thus obtained does not differ from the map determined by other methods. It appears, therefore, possible to relate information on K-12 genetic material gained by genetic and by physical methods.

REFERENCES

ANDERSON, T. F., WOLLMAN, E. L. & JACOB, F. (1957). *Ann. Inst. Pasteur*, **93**, 450.

CAVALLI, L. L. (1950). *Boll. Ist. Sierot. Milano*, **29**, 1.

CAVALLI, L. L. & JINKS, J. L. (1956). *J. Genet.* **54**, 87.

CAVALLI, L. L., LEDERBERG, J. & LEDERBERG, E. M. (1953). *J. Gen. Microbiol.* **8**, 89.

CLOWES, R. & ROWLEY, D. (1954). *J. Gen. Microbiol.* **11**, 250.

FISHER, K. W. (1957). *J. Gen. Microbiol.* **16**, 136.

FREDERICQ, P., & BETZ-BAREAU, M. (1953). *C.R. Soc. Biol., Paris*, **147**, 1653.

FUERST, C. R., JACOB, F. & WOLLMAN, E. L. (1956). *C.R. Acad. Sci., Paris*, **243**, 2162.

FUERST, C. R., JACOB, F. & WOLLMAN, E. L. (1958). *Ann. Inst. Pasteur* (in Preparation).

FUERST, C. R. & STENT, G. S. (1956). *J. Gen. Physiol.* **40**, 73.

HAYES, W. (1953 *a*). *J. Gen. Microbiol.* **8**, 72.

HAYES, W. (1953 *b*). *Cold Spr. Harb. Symp. Quant. Biol.* **18**, 75.

HAYES, W. (1957). *J. Gen. Microbiol.* **16**, 97.

HERSHEY, A. D., KAMEN, M. D., KENNEDY, J. W. & GEST, H. (1951). *J. Gen. Physiol.* **34**, 305.

JACOB, F. & WOLLMAN, E. L. (1955). *C.R. Acad. Sci., Paris*, **240**, 2566.

JACOB, F. & WOLLMAN, E. L. (1956 *a*). *C.R. Acad. Sci., Paris*, **242**, 303.

JACOB, F. & WOLLMAN, E. L. (1956 *b*). *Ann. Inst. Pasteur*, **91**, 486.

JACOB, F. & WOLLMAN, E. L. (1957). *C.R. Acad. Sci., Paris*, **245**, 1840.

LEDERBERG, J. (1947). *Genetics*, **32**, 505.

LEVINTHAL, C. (1954). *Genetics*, **39**, 169.

McCLINTOCK, B. (1956). *Cold Spr. Harb. Symp. Quant. Biol.* **21**, 197.

STENT, G. S. (1953). *Cold Spr. Harb. Symp. Quant. Biol.* **18**, 255.

STENT, G. S. & FUERST, C. R. (1955). *J. Gen. Physiol.* **38**, 441.

TATUM, E. L. & LEDERBERG, J. (1947). *J. Bact.* **53**, 673.

WOLLMAN, E. L. & JACOB, F. (1955). *C.R. Acad. Sci., Paris*, **240**, 2449.

WOLLMAN, E. L. & JACOB, F. (1957). *Ann. Inst. Pasteur*, **93**, 323.

WOLLMAN, E. L., JACOB, F. & HAYES, W. (1956). *Cold Spr. Harb. Symp. Quant. Biol.* **21**, 141.

FERTILITY FACTORS IN *ESCHERICHIA COLI*

By HELEN L. BERNSTEIN

The Lister Institute of Preventive Medicine, London

INTRODUCTION

In a lecture to the Royal Society in 1945, Muller (1947) developed the concept of the gene as 'the dynamic constant' of biological material. Living things, despite their complexities of characteristics and processes, tend to go in cycles, or generations, and at the end of every cycle the starting-point is reached again. This return to the starting-point of a complex course implies a guide, or constant, 'a relatively stable controlling structure to which the rest is attached and about which it in a sense revolves'. Since this return to the beginning finds all structures doubled, or replicated, so it must be also that the material furnishing the frame of reference undergoes reproduction or replication, and this, too, must have occurred under its own guidance. The genetic material, evidently of molecular or macromolecular structure, provides all the information necessary for its own replication from heterogeneous non-genetic substances; hence, it is a dynamic, rather than a static, constant. The integrity of gene stability is inferred from the rare exception—the occurrence of a sudden change or mutation, which is thereafter replicated—again, a 'stable self-multiplying pattern'.

The linear array of genes in chromosomes has ensured the perpetuation of combinations of genes, and the evolution of mechanisms of gene recombination has allowed species possessing such mechanisms to present endless new combinations to meet the selective demands of the conditions in which they live. In addition to crossing over (or something analogous to it) which is the key process in gene recombination, there is also recombination of entire chromosomes in multi-chromosomal organisms. At high levels of organization, one finds physiological or sexual differentiation of nuclei; and at still higher levels differentiation of cells and of entire organisms. These differences all ensure a high probability of fertilization—essential in organisms where sexual reproduction is the only method available.

Gene recombination follows union of the genetic material (genome) of one organism with all or a part of the genome of another. Complete fertilization, resulting from the union of two entire genomes, is found among all sexual forms. Incomplete fertilization (union of one genome with a part of another) followed by gene recombination resulting in the formation of new individuals, is recorded in a few species, and the mechanism of fertilization depends on the species. For example, in the case of *Pneumococcus*, treatment of a culture *A* with highly purified preparations of DNA from a genetically different culture *B* (Avery, MacLeod & McCarty, 1944) leads to the transformation of some cells in the treated culture, and these cells may be shown to have obtained, from culture *A*, genetic material which is permanently incorporated into the genome of the transformed cells. In *Salmonella* species (Zinder & Lederberg, 1952) the partial fertilization is effected by bacteriophage which carry or transduce small fragments of genetic material from one cell to another, where the genes may subsequently be incorporated into the genome.

Escherichia coli is unusual in that fertilization may be either partial or complete. The genetic material of this haploid species appears to consist of a single linkage group or chromosome and is replicated, as in other species, once every nuclear division. But the *factors relating to fertility*— in partial or complete fertilization—sometimes depart from this rule. It is, therefore, of interest to review what we know of their replication.

PARTIAL FERTILIZATIONS

Transduction in *E. coli* was first reported by Morse (1954) who found that the bacteriophage lambda (λ), carried by many cultures of *E. coli* strain K-12, could transfer markers controlling galactose fermentation from one cell to another. These Gal loci are very near to Lp, the locus for lambda maintenance. No other loci could be transduced by lambda, but Lennox (1955), working with phage P1 found that it could transduce almost any genetic marker in K-12, and could also be used for intra-specific transductions. At this time Jacob (1955) also reported transduction of lysogeny in *E. coli*. What appears to be a cell-to-cell transfer is the transmission of colicinogenesis from one strain of *E. coli* to another, reported by Fredericq (1954) and discussed in detail elsewhere in this symposium. These cases all represent partial fertilization of one cell with a small amount of genetic material from another, followed by incorporation of the imported piece into the genome.

SEXUAL RECOMBINATION

The first conclusive demonstration of a sexual process in *E. coli* was reported ten years ago by Tatum & Lederberg (1947). They mixed B-M- (requiring biotin and methionine for growth) and T-L-B$_1$- (requiring threonine, leucine, and thiamine) auxotrophic mutants of *E. coli* strain K-12 in a minimal medium and were able to isolate prototrophs—recombinants possessing none of the parental nutritional requirements, thus resembling the K-12 prototype. Other hereditary factors (such as sugar fermentation and drug resistance markers) not selected by the conditions of crossing, were also found to be recombined in the prototrophs. Early efforts at making genetic maps of these factors on the basis of their recombination frequencies (Lederberg, Lederberg, Zinder & Lively, 1951) were complicated by a puzzling bias of results from cross to cross, which prompted Muller (in a discussion appended to this reference) to suggest that the parents might be of different mating types. K-12 was thought at that time to be homothallic; soon, however, physiological differences between parent cultures were reported by Hayes (1952*a, b*), and a mating-type system was disclosed by the Lederbergs and Cavalli (1952).

Certain M- mutants of 58–161 (the original B-M- parent which had reverted to B$^+$) were found to be infertile with T-L-B$_1$ (that is, incapable of mating to produce prototrophic recombinants), although 58–161 remained fertile. If the 58–161 M-* derivatives were allowed contact with 58–161, and were then re-isolated and tested with T-L-B$_1$, they were found to be fertile, and this fertility persisted through all subsequent generations. Similarly, if T-L-B$_1$- cultures were allowed prior contact with 58–161, they became permanently fertile with the M- mutants. The fertility of 58–161 was attributed to the presence of a hypothetical infective agent, *F*, which is transmissible by cell contact only and has never been isolated in a cell-free state. Cultures having this transmissible fertility were termed *F*$^+$. The infertile cultures mentioned above lacked *F* and were called *F*$^-$. When these cultures were made *F*$^+$ by contact with *F*$^+$, they became stable *F*$^+$ cultures and were in turn able to transmit *F* to other *F*$^-$ cultures. K-12 itself was found to be *F*$^+$; hence, the *F*$^-$ mating types must have arisen independently in the two lines. The progeny of all *F*$^+ \times F^+$ and *F*$^+ \times F^-$ crosses are reportedly *F*$^+$ (Lederberg *et al.* 1952); Hayes, however (1953) reports that while the progeny of M-*F*$^+ \times$ T-L-B$_1$-*F*$^-$ are all *F*$^+$, those of T-L-B$_1$-*F*$^+ \times$ M-*F*$^-$ are mixed, *F*$^+$ and *F*$^-$, the inheritance of *F* not being linked to that of any other genetic marker. This segregation of fertility has been confirmed (Bernstein, 1957).

A third sexual or mating type was found by Cavalli (1950) as a mutation of 58–161 to an extremely fertile state; because of its high frequency of recombination it was designated *Hfr*. Unlike *F*$^+$, *Hfr* has no transmissible agent; yet unlike *F*$^-$, it cannot acquire a transmissible agent by contact with *F*$^+$. Progeny of *Hfr* $\times$ *F*$^-$ are mixed, *Hfr* and *F*$^-$. *Hfr* mating type appears

---

* Hereafter, when M- and T-L-B$_1$- mutants are mentioned, it is always with reference to these two mutant lines of K-12.

to be genetically controlled, and is closely linked in the chromosome to the loci controlling galactose fermentation. Another M-*Hfr* mutant was isolated by Hayes (1953); this also lacked a transmissible agent, could not be infected by contact with F^+, and appeared to be genetically determined. Its high fertility was only expressed in crosses to T-L-B_1- when B_1 was included in the medium (i.e. when no selection for B_1 was employed). Occasionally Hayes' *Hfr* culture back-mutates to F^+, having correspondingly lower fertility and again being able to transmit *F*. Recently Richter (1957) has reported a third *Hfr* mutant '*Hfr* 3' of high fertility and not transmitting *F*. *Hfr* 3 back-mutates at a high frequency to F^+. Cavalli's, Hayes' and Richter's *Hfr* genetic determinants are all at different loci. Progeny from the cross *Hfr* 3 × F^- include F^+, *Hfr* 3 and two kinds of F^-. Of the latter, one, after *F* infection, becomes F^+, while the other (a very rare class) becomes *Hfr* 3. These findings were interpreted as meaning that *F* has a special affinity for the *Hfr* 3 locus and may become stably fixed, or bound to this locus. (Such a mechanism of bound *F* had previously been considered as an explanation for the other two *Hfr* mutants.)

In the Lederberg laboratory (Lederberg *et al.* 1952) a search for the prevalence of occurrence of *F* was made by crossing about 2000 strains of *E. coli* to a T-L-B_1-F- culture (W1177), using streptomycin selection. About forty were fertile and were presumed to be F^+. (Seven of these, tested for the presence of *F*, were able to make K-12 F^- fertile by contact, but it was noted that foreign *F* was often unstable and that the infected cells reverted to F^-.) By retesting some of the strains with a K-12 F^+ tester, another group similar in number were found to be fertile with F^+, though not with the F^- tester. These were presumed therefore to be F^-.

MATING TYPES

The mating types in the K-12 system differ from each other in three major respects:

(1) In their mating behaviour with cultures of the same or different mating type.

(2) In their ability to alter, by cell contact, the mating type of cells of other mating type—the possession or lack of transmissible fertility *F*.

(3) In their ability to undergo such alteration of mating type by cell contact—susceptibility to infection by *F*.

A further consideration is the mechanism of determination of the mating type, reflected by the inheritance of mating type in various crosses.

To approach a complete understanding of the basis of mating type and fertility in *E. coli* we must make use of all these criteria, especially in any search for types differing from those in K-12. Mating behaviour is judged not only by the degree of fertility, but also by the frequencies of gene recombination among the prototrophic progeny. Transmission of mating type is detected by incubating mixed cultures of different types in broth, re-isolating the one to be tested, and then re-examining it by the above criteria. In assessing mating type, one can obtain both qualitative and quantitative information about each criterion.

Because of the numerous traits characterizing a mating type and because (as it has turned out) of the sexual diversity in this species, there is no simple system of nomenclature for mating types. F^+ and F^- were named according to the presence or absence of F; Hfr was so named because of its high frequency of recombination. However, because of the additional characteristics of these types, it is confusing to use the same terms to designate other types which may in fact have only a single characteristic in common with the original type of this designation. It is hoped that these semantic difficulties may some day be surmounted, but for purposes of the present discussion mating types will be referred to in terms of the above criteria and according to the strain or culture in which they are detected.

Employing the criteria of mating behaviour, possession of F, and susceptibility to infection by F, Calef & Cavalli-Sforza (1954) reported the existence of a new mating type in *E. coli* strain B. This strain is fertile with K-12 F^+ and Hfr, and infertile with F^-. It thus resembled F^- cultures. However, it could not be made fertile with F^- by contact with F^+ (see Table 1). In a cross of T-L-B$_1$-F^+ by a B tryptophaneless mutant, the progeny were nearly all F^-, only one F^+ being found among several hundred prototrophs. This F^+ hybrid was fertile with, and could transmit F to, K-12 F^- cultures. In crosses between this hybrid and K-12 F^-, the fertility segregated among the progeny, no linkage to other markers being reported. Crosses between Hfr and B yielded rare fertile prototrophs which (like their Hfr parent) were not able to transmit F to F^- cultures.

Recently I have investigated several strains of *E. coli* in a search for new mating types (Bernstein, 1956), feeling that the more mating types there are available for study, the greater are our chances for understanding the basic factors affecting fertility in this species. Details of these investigations are being published elsewhere, but the results are summarized in the following paragraphs and in Table 1. Nearly every strain examined revealed something new, and none of them was identical to K-12 mating types.

Two strains (WG54 and WG55) resembled *E. coli* B. They mated with K-12 F^+ but not with F^-, and efforts to make them fertile with F^- were unsuccessful. Another strain (WG53) was similar to F^- in its mating behaviour, and could be made fertile by contact with F^+. However, auxotrophic mutants of this strain, while infertile with K-12 F^-, were fertile with each other. Strain WG52 also appeared at first to be F^-: it mated with F^+, not with F^-, and could be infected by F, after which it was fertile with F^-. But in spite of this alteration of fertility, the infected culture did not transmit F to K-12 F^- cultures.

Another strain (WG4) was studied in considerable detail. It was fertile in crosses with M-F^+ and T-L-B$_1$-F^+ and M-F^-, but not with T-L-B$_1$-F^-. (If the latter was made F^+, it could then mate with WG4.) Mixtures were made of WG4 with each of the K-12 F^- testers, but F was not transferred to either of them. This strain, then, resembles *Hfr* in being fertile without possessing a transmissible agent. But unlike *Hfr*, WG4 can easily be

Table 1. *Mating types in* Escherichia coli

Strains were studied for their fertility with K-12 testers, for their possession of transmissible fertility (F), and for their susceptibility to infection by the F from K-12 ($F12$).

Strains of E. coli	Fertility with K-12 F^-		Possession of a transmissible agent (F)	Susceptibility to infection by $F12$	Other criteria
	M-	T-L-B$_1$-			
K-12 F^-	—	—	—	+	—
K-12 $F12^+$	+	+	+	o	—
K-12 *Hfr*	+	+	—	—	—
B WG54 WG55	—	—	—	—	—
WG52	—	—	—	+	—
WG52 $F12^+$	+	+	—	o	—
WG53	—	—	—	+	Self-fertile
WG4	+	—	—	+	Self-sterile
WG4 $F12^+$	+	+	+	o	$F12$ unstable
WG3	+	+	+	$(+)^*$	$F3$ differs from $F12$

+, −, indicate the presence, or absence, of the various criteria; o, no test made.
* See text.

infected and made F^+: it is now fertile with T-L-B$_1$-F^- as well as M-F^-, and it can transmit F to K-12 F^- cells. The F from K-12 is only moderately stable in WG4, and WG4 'F^+' reverts to the original WG4 mating type. The progeny of WG4 × M-F^- were of either F^- or WG4 mating type. The segregation of WG4 mating type as a unit in this cross suggests it is genetically determined. Further support for this was obtained when a cross of WG4 × M-*Hfr* yielded an F^- recombinant, which is what would be expected from a cross of two mating types determined by non-allelic genes. Crosses of two WG4 mutants with each other were infertile, which indicates that this strain may be self-sterile.

The fertility of WG4 with one F^- but not the other, suggested that there was a difference between the M- and T-L-B$_1$- lines of K-12 which was not attributable to nutritional selection (since WG4 mated with T-L-B$_1$-F^+). Hybrids between the M- and T-L-B$_1$- lines, made by $Hfr \times F^-$ crosses, showed that this difference segregated among the progeny independently of Hfr. It was found, however, that the Hfr gene was epistatic to that controlling compatibility with WG4, in that an Hfr carrying the factor for incompatibility with WG4 is nevertheless fertile with WG4. This is in agreement with the observation that T-L-B$_1$- made F^+ is fertile with WG4.

The last of this group to be investigated was WG3, which had an F agent and was fertile with F^+ and F^-, although quantitative differences in fertility distinguished it from K-12 F^+. This strain could transmit F to M-F^- but not to T-L-B$_1$-F^-. However, the infected M- could transmit the WG3 agent to T-L-B$_1$-F^-, and the latter, when infected, could transmit the agent to other F^-. When a culture of WG3 was allowed contact with a culture of K-12 F^+, WG3 became able to infect T-L-B$_1$-F^-. Whether this represented a state of mixed F infection, in WG3, or whether the F from K-12 replaced the F of WG3, is not yet known.

In addition to the types shown in the table, Lederberg & Lederberg (1956) mention strains which are sterile with F^- (T-L-B$_1$-F^-) but have F; a strain which is fertile with T-L-B$_1$-F^- and has no F but can be infected; and another, also fertile with T-L-B$_1$-F^- and having no F, but not infectable.

If one considers the criteria (fertility with a K-12 F^-, possession of a transmissible fertility factor F, and susceptibility to infection by contact with a K-12 F^+) from only a qualitative point of view, there would be eight possible mating types. Examples of each one exist among the strains described above. Further types are theoretically possible by quantitative considerations of these criteria, and by broadening the reference (i.e. fertility with M-F^-, T-L-B$_1$-F^-). A chessboard study of the fertility of each mating type known with that of every other type would presumably give us still further new criteria for the assessing of mating type. However, this study was stopped in order to investigate in greater detail the subject of transmissible fertility.

TRANSMISSIBLE FERTILITY

The investigation of WG3 suggested that the F native to it might differ from the F of K-12, and it was decided to compare these agents as fully as possible (Bernstein, 1957). The F of WG3 is referred to as $F3$, and that of K-12 as $F12$. By starting with M-F^- and T-L-B$_1$-F^-, it was possible to have F^-, $F3^+$, and $F12^+$ in each of these lines of K-12, allowing full comparisons of the agents in a uniform genetic background. $F3$ and $F12$ were found to differ (1) in efficiency of transmission from F^+ to F^-, (2) in the degree of fertility conferred upon infected cells, (3) in their influence on recombination frequencies in various regions of the genome, and (4) in their inheritance among the progeny of $F^+ \times F^-$ crosses.

When F^+ and F^- cultures are mixed in a complete medium, the F^- cells become F^+ by contact with F^+ cells. This does not occur in a minimal medium. F_{12} not only differs from F_3 in its efficiency of transmission (frequency with which F^- becomes F^+), but the efficiency of F_{12} transmission also depends on the parent bearing it in the first place. When M-F^+ cells were mixed with T-L-B_1-F^-, and the latter were re-isolated after 3 hr., 70–100% of the F^- became F^+ if the M- was F_{12}^+, but only 10–20% if the M- was F_3^+. When T-L-B_1-F^+ cells were mixed with M-F^-, only 10–20% of the M- became fertile after 3 hr., whether the T-L-B_1- was F_{12}^+ or F_3^+. After overnight incubation of the mixtures, the F^- components of all mixtures had been converted to F^+. From all these mixtures, after both hourly and overnight incubation, re-isolates of the F^+ components when retested were still F^+; and viable counts showed no decrease in absolute numbers of either M- or T-L-B_1-; hence, whatever the agent of transmissible fertility, the supply was not exhausted by the increased distribution.

The fertility of F_{12}^+ and F_3^+ cultures differs. In $F^+ \times F^-$ crosses the $F_3^+ \times F^-$ are only about half as fertile as $F_{12}^+ \times F^-$, when crossed on minimal medium. $F_{12}^+ \times F_{12}^+$ are as fertile as $F_{12}^+ \times F^-$, but $F_3^+ \times F_3^+$ are of very low fertility. The most fertile crosses are $F_3^+ \times F_{12}^+$, these being at least 2–4 times as fertile as $F_{12}^+ \times F^-$.

The study of segregation of unselected markers in all the crosses showed that recombination frequencies in various regions of the genome depended not only on whether a parent was F^+ or F^-, but also on which F was involved. The progeny of M-$F_{12}^+ \times$ T-L-B_1-F^- tended to get most of their markers from the T-L-B_1-F^- parent. But the progeny of all other crosses (T-L-B_1-$F_{12}^+ \times$ M-F^-, both $F_3^+ \times F^-$, and all $F^+ \times F^+$) got most of their markers from the M- parent. In all regions of the genome which were studied, the effects of the two F agents on frequencies of gene recombination were different.

The inheritance of F among the progeny of $F^+ \times F^-$ crosses has already been remarked on in the case of F_{12}^+; all progeny of M-$F_{12}^+ \times$ T-L-B_1-F^- are F^+, while progeny of T-L-B_1-$F_{12}^+ \times$ M-F^- are either F^+ or F^-. The progeny of $F_3^+ \times F^-$ crosses are also mixed, F^+ and F^-. The conditions of crossing (on minimal medium) do not favour transmission of F by cell contact, and in all these crosses but one, mating type is segregating. Mating-type determination was not linked to any other unselected marker in any of these crosses.

DISCUSSION

The mating type of a strain or culture of *Escherichia coli* is, like its other characteristics, stable, though rare permanent changes or mutations can and sometimes do occur. This inherent stability of a complex pattern of behaviour suggests genetic control, and in certain strains genetic determination of mating type has been demonstrated. Recombination between genes for mating type has shown that more than one locus is involved. These genes do not all act in the same way, for a factor making a strain fertile with some cultures will not necessarily make it fertile with others. Effective cell contact, gene transfer, and gene recombination comprise a complex process, and presumably different genes could affect different aspects of this process. Thus, one mating type is fertile with other mating types though self-sterile, while another is self-fertile though sterile with other apparently analogous types. A single cell may possess more than one mating-type gene, and one such gene may be epistatic to another. Two cultures hitherto thought to be of the same mating type (M-F^- and T-L-B_1-F^-) have been found to differ in their possession of a factor controlling compatibility with another mating type (WG4). It is clear that as our available criteria of mating type expand, we shall be able to disclose more and more differences among mating types.

Certain strains of *E. coli* are able to transmit fertility to other strains by cell-to-cell contact. When this occurs the mating type of the recipient strain is altered in some way, while that of the donor strain remains unaltered. The speed and efficiency with which this transmission can occur argues that F (whatever is being transmitted) is present in excess of the other genetic material of the cells carrying it. Different efficiencies of transmission of F_{12} and F_3 from M-F^+ to T-L-B_1-F^- may be due to different effects that the agents have on the ability of the F^+ cell to transmit F, or it may be that these agents are maintained at different levels in this F^+, with correspondingly different probabilities of transfer. Possibly, both hypotheses are true. The higher efficiencies of infection obtained with M-F_{12}^+ compared with those with T-L-B_1-F_{12}^+ suggest that the same F may be replicated at a higher frequency in one culture than in another. The tendency for certain infected strains to lose F indicates that F is replicated at too low a level in these strains and is eventually diluted out. At times F may become bound, and be no longer transmissible, but be replicated at the same rate as the rest of the genome, and undergo segregation and recombination with other markers. This binding may occur in an F^+ culture, giving an *Hfr* derivative, or it may follow F infection of certain cultures. In *Hfr 3* the binding appears to consist of an association between F and the *Hfr 3* locus in the chromosome, and it is not unreasonable to suppose that this may be the mechanism at least in the case of the other *Hfr*'s. Whether there is some such locus related to *F* maintenance in the *F+* is not known; none has been detected in recombination tests but this may be due to the limitation of the selective techniques employed in crosses.

From studies of *F3* and *F12* it was found that different *F* agents have different effects on fertility and on frequencies of recombination in various regions of the genome. (It may be that the fertility differences are simply one result of the *F* effect on recombination frequency.) The capacity of *F* to affect these frequencies is of further interest when one considers that in two of the *Hfr* mutants, both arising from *F12+* cultures, the recombination frequencies also differ. If the concept of *F* being bound to the *Hfr* locus applies to these two *Hfr*'s, then it would appear that there is a sort of position effect on *F*, the recombination frequencies depending on where it gets bound.

One question of great interest is whether or not one can have simultaneous infection with two *F* agents. In the case of WG3, there is a suggestion that this may be so. However, it might instead have been that *F12* replaced *F3*, or that some cells were *F3+* and some *F12+*. The study of mixed infection with two transmissible agents is made extremely difficult by the lack of simple diagnostic techniques for distinguishing the two agents. The case of WG4 is a rather different one. This strain is fertile though lacking a transmissible agent, and its mating type is genetically determined. If it were postulated that WG4 fertility, like that of *Hfr 3*, was due to the binding of an *F* agent at a locus, then WG4, which had been made *F12+* by contact with K-12, would be a clear case of mixed infection, with *F12* having no affinity for the WG4 mating-type locus.

While *F*, transmissible fertility, has much in common with more orthodox genetic factors affecting gene recombination, it also has something in common with bacteriophage in that it is replicated at a higher frequency or maintained at a higher level than other genes of the cell. Both *F* and bacteriophage are intimately related with the genetic material of the cell. Just as some strains are resistant to phage infection, so some appear to be resistant to *F* infection. But *F* differs from phage in not being separable from the cell.

F also bears certain resemblances to colicinogenic factors, in being transmissible by cell contact, and, so far as is known, in having no linkage to other genes of the cell. Both are replicated at higher levels than other genes. But, while similar in their replication and transmissibility, they are markedly different in their actions. Indeed, their end-effects on the likelihood of gene recombination between different cultures are nearly opposite.

The presence of *F* tends to increase this likelihood, while the elaboration of a colicin definitely decreases it.

In *E. coli* the 'dynamic constants' determining fertility are sometimes localized in the chromosome, segregating with other genes and being replicated at the same frequency as other genes. In other cases mating-type determinants are transmissible from cell to cell, showing no linkage with the chromosome and evidently being replicated in excess of other genetic material. Occasionally such a determinant becomes 'bound' in the cell in such a way that although transmissibility is lost, fertility is retained (or even increased), and thereafter it is seemingly replicated at the same rate as the rest of the genome. Not only do the factors determining fertility appear to be genetic in nature, but there are indications that the rate at which such factors are replicated is also under genetic control.

REFERENCES

AVERY, O. T., MacLEOD, C. M. & McCARTY, M. (1944). *J. Exp. Med.* **79**, 137.

BERNSTEIN, H. L. (1956). *Abstr. Genet. Soc.* in *Heredity* (1957), **11**, 154.

BERNSTEIN, H. L. (1957). *J. Gen. Microbiol.* **16**, iii.

CALEF, E. & CAVALLI-SFORZA, L. L. (1954). *Microb. Gen. Bull.* **10**, 8.

CAVALLI, L. L. (1950). *Boll. Ist. Sieroter. Milano*, **29**, 281.

FREDERICQ, P. (1954). *C.R. Soc. Biol., Paris*, **148**, 399.

HAYES, W. (1952*a*). *Nature, Lond.* **169**, 118.

HAYES, W. (1952*b*). *Nature, Lond.* **169**, 1017.

HAYES, W. (1953). *Cold Spr. Harb. Symp. Quant. Biol.* **18**, 75.

JACOB, F. (1955). *Virology*, **1**, 207.

LEDERBERG, J., CAVALLI, L. L. & LEDERBERG, E. M. (1952). *Genetics*, **37**, 720.

LEDERBERG, J. & LEDERBERG, E. M. (1956). *Cellular Mechanisms in Differentiation and Growth*. V. Infection and Heredity. Princeton University Press.

LEDERBERG, J., LEDERBERG, E. M., ZINDER, N. D. & LIVELY, E. R. (1951). *Cold Spr. Harb. Symp. Quant. Biol.* **16**, 413.

LENNOX, E. S. (1955). *Virology*, **1**, 190.

MORSE, M. L. (1954). *Genetics*, **39**, 984.

MULLER, H. J. (1947). *Proc. Roy. Soc. B*, **134**, 1.

RICHTER, A. (1957). *Rec. Gen. Soc. Amer.* **26**, 391.

TATUM, E. L. & LEDERBERG, J. (1947). *J. Bact.* **53**, 673.

ZINDER, N. & LEDERBERG, J. (1952). *J. Bact.* **64**, 679.

COLICINS AND COLICINOGENIC FACTORS

By PIERRE FREDERICQ

University of Liége

I. COLICINS

The large intestine of man and lower mammals is a veritable culture tube in which definite bacterial types appear to be struggling constantly to gain supremacy (Rettger, 1928). To that end they use all the known, and unknown, mechanisms of antibiosis, and antagonistic manifestations due to *Escherichia coli* or other enteric bacteria have often been reported in the literature (cf. Fredericq, 1948e; Parmala, 1956).

Antagonistic manifestations due to the production of a diffusible antibiotic substance were, however, first demonstrated by Gratia (1925). A strain V of *E. coli* inhibited the growth of another strain φ of the same species, as well as a *Shigella dysenteriae*, but had no action upon some other *Escherichia coli* strains tested. It produced a highly potent antibiotic substance, named 'principle V', which was still active in dilution 1/1,000, could diffuse in agar, and pass through cellophane membranes, was precipitated by acetone but was not antigenic (Gratia, 1932).

After some further isolated reports (Guelin, 1943; Zamenhof, 1945) of the production of an antibiotic substance by *E. coli* or related organisms, a systematic investigation (Fredericq, 1946c; 1948e) revealed that antibiosis of the type coli V—coli φ is of much more frequent occurrence and less specific than was at first thought. Among the many strains tested, about 25 % inhibit the growth of strain *E. coli* φ and still more (about 50 %) are inhibited by the strain *E. coli* V. If the strains which are active against coli φ are now tested against many other strains, they display strikingly different spectra and must therefore produce a number of quite different antibiotic substances which were designated *colicins*. There are colicin A, B, C ... and principle V of Gratia changes quite naturally into colicin V.

Colicinogenic strains

Colicinogenic strains, that is strains which produce colicin, may be searched for by the following technique: The strains to be tested are stabbed on an ordinary agar plate (generally 8 per plate) and incubated for 48 hr. The macrocolonies which develop are sterilized by chloroform vapour. The sensitive indicator strain is then seeded over the whole surface, either by pouring on to it 4–5 ml. of melted agar containing 0·1 ml.

of a broth culture of that strain, or by applying for 5 min. a disk of filter-paper soaked with the culture. After 24 hr. incubation, the confluent growth of the indicator reveals inhibition zones surrounding the colicinogenic colonies. The susceptible indicator strain may be *E. coli* φ of Gratia or the well-known *E. coli* B or K-12, which seem to be susceptible to all colicins. It is, of course, possible that they are not susceptible to all colicins, but experiments with a limited number of other strains never revealed an antibiotic action which would have been missed with the above indicators.

Colicinogenic properties are quite frequent in *E. coli* strains (Fredericq, 1946c; Halbert & Gravatt, 1949b; Mondolfo, 1948). Many related strains such as paracoli or *E. freundii* (Fredericq, 1948e), but also *Shigella* (Fredericq, 1948a; Fastier, 1949) and even, though more rarely, *Salmonella* (Fredericq, 1952a; Hamon, 1955) produce antibiotic substances which are identical with or related to those produced by *E. coli* and must therefore be included in the colicin group.

Colicinogenic strains have been reported from many parts of the world (Fredericq & Levine, 1947; Halbert, 1948a; Mondolfo, 1949; Chabbert, 1950; Grosso, 1950; Blackford, Parr & Robbins, 1951; Ludford & Lederer, 1953; Levine & Tanimoto, 1954). The use of selective techniques revealed their presence in nearly all human and animal stools tested (Gratia, Fredericq, Joiris, Betz-Bareau & Weerts, 1950). They appear in the first days of life and are found in individuals of all ages.

Colicinogenic strains act in various ways, producing many different colicins, whose distinctive characters are enumerated in the next section. Some strains may release at the same time two or more distinct colicins. Colicinogenic strains are not susceptible to the particular colicin they produce but may be inhibited, just like non-colicinogenic strains, by other colicins, and manifest intricate patterns of reciprocal antagonism (Fredericq, 1948e).

Colicinogenic strains do not differ from non-colicinogenic strains by any other properties, except that they often yield mucoid colonies (Mondolfo & Ceppellini, 1950). There seems, however, to be some relationship between biochemical and serological properties of the strains and the type of colicin they produce (Fredericq, 1946a; 1948e). For example, *E. coli* strains may produce all the types but most of them release colicin V. Most paracoli strains produce colicins E or K, most *Shigella sonnei* colicin E, and *Escherichia freundii* colicin A. All analysed colicinogenic *Salmonella* strains, including three *S. typhi-murium*, one *S. para* B and one *S. bareilly*, were found to yield colicin I.

There seems also to exist a relationship between colicinogenic strains and intestinal affections (Fredericq, 1953b). Strains producing colicins which are active against *Shigella* are more frequently found in dysenteric patients than in normal individuals (Halbert, 1948*b*, *c*). Strains producing colicin B, one of the rare colicins active against *S. para* B, are more frequent in paratyphoid B patients (Fredericq, Joiris, Betz-Bareau & Gratia, 1949). A special type of *Escherichia freundii*, producing colicin A, is found regularly and consistently in the urine of typhoid patients, even when it is aseptically collected (Fredericq & Betz-Bareau, 1948). Furthermore, so-called pathogenic *E. coli* strains isolated from infantile gastro-enteritis are often colicinogenic and most of them produce the colicins E and I characteristic of *Shigella* and *Salmonella* (Fredericq, Betz-Bareau & Nicolle, 1956). Indeed, the colicinogenic properties of pathogenic *Escherichia coli*, as well as of *Shigella sonnei* (Abbott & Shannon, 1956), may be used as a marker in epidemiological studies. This colicin typing is independent of the phage typing and allows a subdivision of the types. The production of colicins by some intestinal pathogens must certainly play a part in their struggle to supplant the normal intestinal flora and may therefore be considered as a factor of pathogenicity.

Nature of colicins

Colicins form a group of antibiotic substances which differ by many properties:

(1) *Extent and specificity of their activity spectra.* Colicinogenic strains have strikingly different activity spectra which are, however, strictly limited to the Enterobacteriaceae family. An action of colicinogenic strains upon bacteria which do not belong to this family, as described by some (Cook, Blackford, Robbins & Parr, 1953), must probably be due to factors other than colicins. Some colicinogenic strains have a very wide range of activity, others have a much narrower one. All of them are active against some strains like ϕ, B or K-12, but many other strains are resistant to one or more and show the most varied patterns. Susceptibility-patterns may be used for typing the *Coli-Shigella-Salmonella* group (Fredericq, 1948*e*; Halbert & Gravatt, 1949*a*; Robbins, El Shawi & Parr, 1956) and seem to have a connexion with the antigenic constitution of the strains. *E. coli* and *Shigella* strains are often sensitive to many different colicins, *Aerobacter* and *Salmonella* are less frequently susceptible and then to a limited number only. Activity spectra of colicins are highly specific and diverse, and quite comparable to those of bacteriophages (Fredericq, 1950*e*; 1953*a*).

(2) *Specificity of resistant mutants.* In the inhibition zones produced on an indicator strain by colicinogenic strains, colonies of resistant mutants often arise. These mutants are no longer susceptible to the colicinogenic strain in whose inhibition zone they appeared, nor to other colicinogenic

strains producing the same colicin. They are still inhibited and remain completely susceptible to strains producing other colicins (Fredericq, 1946 b).

(3) *Diffusibility.* Colicins do not diffuse rapidly in agar and colicinogenic strains have to be stabbed and incubated during 48 hr. before applying the indicator strain, in order to show well-defined inhibition zones. The rate of diffusion of each colicin is, however, very different. Colicinogenic strains producing a highly diffusible colicin are surrounded by a wide inhibition zone, up to 40 and even 50 mm. in diameter; others have a much narrower one. Highly diffusible colicins, like colicin V, are able to pass through cellophane membranes but most other colicins are kept back (Fredericq, 1948 e).

(4) *Morphology of the inhibition zones.* Beside their size, which is largely due to diffusibility of the colicins, the inhibition zones surrounding colicinogenic strains differ widely in their morphological aspect. They may be without any growth, reveal a continuous partial growth or a variable number of resistant colonies. Their edge may be clear-cut or gradual, sometimes rosette-shaped, and surrounded later with a zone of secondary lysis. The varied appearances of the inhibition zones reproduce in fact, to a much bigger scale, every morphological modality of the plaques produced by bacteriophages. In both cases they depend on the specificity of the inhibiting agent and on the nature of the inhibited strain (Fredericq, 1948 e).

(5) *Susceptibility to proteolytic enzymes.* Colicins are more or less rapidly destroyed by proteolytic enzymes such as trypsin as well as those produced by many strains of the genera *Bacillus, Micrococcus, Enterococcus, Pseudomonas* and *Proteus.* Their different susceptibility may be demonstrated by the following technique: A streak of a colicinogenic strain and, perpendicular to it, a streak of a proteolytic strain are made on an ordinary agar plate, and incubated for 48 hr. The cultures which develop are killed by chloroform vapour, and the whole surface is then inoculated with an indicator strain. Next day it will be seen that the inhibition zone which should regularly surround the colicinogenic streak is destroyed in the vicinity of the proteolytic streak, following a curve whose slope is characteristic for each colicin (Fredericq, 1948 b).

(6) *Thermoresistance.* Colicin V of Gratia was able to withstand heating at 100° C., but other colicins are altered or even completely destroyed at 60–70° C. (Fredericq, 1948 e).

(7) *Electrophoretic motility.* It has been possible to show electrophoretically that some colicins possess a negative, others a positive charge. They may be divided into four groups according to their electrophoretic motility (Ludford & Lederer, 1953).

(8) *Antigenic properties.* Colicins may be differentiated by their antigenic properties, and sera which specifically neutralize one colicin have
been obtained. Colicins are not, however, highly antigenic substances
(Bordet, 1947; Goebel, Barry & Shedlovsky, 1956; Hamon, 1956*a*,
1957*b*).

On the basis of the above-mentioned differences, colicins produced by
various strains may be analysed and separated. During a preliminary
research carried out on a limited number of colicinogenic strains, seventeen
different colicins were demonstrated (Fredericq, 1948*e*). As this number
was likely to increase considerably if more strains were investigated, it was
necessary to adopt a more simple classification. Colicins are now grouped
according only to the specificity of their resistant mutants and all colicins
which select the same type of resistant mutant are put in the same group
(Fredericq, 1953*b*). The colicins of each group have the same specificity of
action but may be very different substances chemically. As a rule mutants
are only resistant to the colicins of one group but sometimes there arises
partial or complete cross-resistance with colicins of other groups, particularly in the case of group B, I and V.

Analysis of the colicins produced by unknown strains is, therefore, not
always easy. Their action has to be tested against a number of mutants,
resistant to different known colicins, in order to determine their specificity.
Alternatively a mutant, resistant to the action of the unknown strain, may
be derived and tested against standard colicinogenic strains.

Analysis of the colicins produced by unknown strains is furthermore
complicated by the fact that many strains produce at the same time two or
even more different colicins. Their activity spectra will represent the sum
of their constitutive colicins but the morphology of the inhibition zone
may be different from that of any one of them. Very often, however, the
two colicins have different rates of diffusibility. The inhibition zone then
presents an external part where only one of the colicins has diffused, surrounding an internal part where both colicins are present. Resistant
colonies may appear in the external part but never in the internal part, as
a double mutation, occurring in the same cell, is only rarely to be expected.
These mutants, resistant to the outer colicin, may be used as indicator to
reveal the second colicin of a double colicinogenic strain. The destruction
curve by bacterial proteases reveals the production of two distinct colicins
when the most diffusible one is also the most susceptible to proteolytic
action. The curve is characteristic of the most diffusible one in the external
part, but it will take the slope of the other one in the internal part as it
occurs in an area where the first colicin, being more susceptible, is already
destroyed (Fredericq, 1948*e*).

Despite their very different properties, all colicins appear to be of a protein nature. They are antigenic substances which are destroyed by proteolytic enzymes. Attempts at purification of colicin V and some others gave products having the general reactions of polypeptides or proteins (Heatley & Florey, 1946; Halbert & Magnuson, 1948; Gardner, 1950; Depoux & Chabbert, 1953).

Colicin K was recently purified by Goebel, Barry, Jesaitis & Miller (1955). After repeated precipitations with ethanol and ammonium sulphate, followed by extraction with chloroform-octyl alcohol, they obtained a colourless, water-soluble substance, which contains 6·5 % nitrogen and 1·6 % phosphorus, but is free from nucleic acid. It displays only one component in electrophoresis but seems to be heterodisperse in the ultra-centrifugation. It has a powerful antibiotic action, for one drop of a solution containing 1 μg./ml. totally inhibits the growth of the indicator strain in agar. It appears to be a macromolecular substance, consisting of carbo-hydrate, protein and lipid and having all the chemical, physical, immuno-logical and toxic properties of the O antigen of the producing bacteria. However, the possibility that colicin K may be a distinct molecule linked to this lipocarbohydrate-protein complex is not excluded. It could well be that the product obtained by Goebel et al. is a mixture of colicin K and its receptor as the receptor seems to be related to the O antigen. According to inactivation-curves by X-rays, colicin K should have a molecular weight somewhere between 60,000 and 90,000 (Latarjet & Fredericq, 1955).

Mode of action of colicins

A strain which is sensitive to several different colicins does not present a single point of attack, characteristic of its type of sensitivity and identical for all colicins to which it is sensitive. Indeed, a resistant mutant derived from a strain susceptible to several colicins does not lose the general sensitivity of that strain at one step, but only the sensitivity to the particular colicin towards which it was directed. It keeps exactly the same suscepti-bility towards other colicins as the strain from which it came. Strains susceptible to several colicins have in fact a series of receptors specific for each colicin to which they are susceptible (Fredericq, 1946b).

These colicin receptors are fixation receptors (Bordet & Beumer, 1951; Fredericq, 1952b) and behave exactly like bacteriophage receptors. They may be extracted from sensitive bacteria and specifically neutralize the corresponding colicin by fixing it. Similar extracts prepared from resistant bacteria have no action (Bordet & Beumer, 1948). Antibacterial sera, which do not have any direct anti-colicin action, protect sensitive bacteria against the later action of colicins, probably by blocking the receptors (Bordet, 1948).

These receptors may be lost by mutations occurring spontaneously (Fredericq, 1948 c). These mutations are specific for each receptor and independent of those affecting other properties (Fredericq, 1948 d). Just as a mutation affecting the receptor for one colicin is independent of the mutation affecting the receptor for another colicin, such mutations are also independent of those affecting bacteriophage receptors. In some cases, however, a specific reciprocal cross-resistance between a colicin and a definite bacteriophage is constantly observed and leads to the conclusion that a single receptor is shared by that colicin and the corresponding bacteriophage. Thus, all resistant mutants, either selected by colicin K or by phage T6, are always resistant to both agents simultaneously (Fredericq & Gratia, 1950). The same holds true with colicin E and phage BF23 (Fredericq, 1949) and with colicin M and phage T1–T5 (Fredericq, 1951).

A mutation brings about the loss of only one receptor at a time, but a mutant resistant to a first colicin may still mutate and become resistant to a second one. Double resistant mutants are useful in analysing the colicins produced by doubly colicinogenic strains. Such resistant mutants may still further mutate, and a strain susceptible to many different colicins may be transformed into a completely resistant one if all of its receptors are taken away one by one in a series of successive mutations (Fredericq, 1948 d).

The presence of genetic factors controlling colicin receptors has to be inferred from the mutation experiments and is also confirmed by recombination studies (Fredericq & Betz-Bareau, 1952). Mutants resistant to colicins, derived from the fertile *E. coli* K-12 strain, are easy to get, and crosses between susceptible and resistant parents were studied, using Lederberg's technique (1947).

Susceptibility to colicins behaves in these crosses like any other genetic marker, in accordance to the general rules of recombination (cf. Lederberg, 1955). Susceptibility or resistance to a given colicin are allelic characters which are linked to other markers and segregate with them according to the selected markers and to the F-polarity of the parents. Let us take, for example, a cross where the F^+ parent is methionine-dependent but threonine-, leucine- and thiamine- independent ($M^-TLB_1{}^+$) and resistant to colicins E, K and V, and where the F^- parent is conversely $M^+TLB_1{}^+$ and susceptible to colicins E, K and V. Most of the prototrophic recombinants ($M^+TLB_1{}^+$) are susceptible to colicin V like the F^- parent, because the marker susceptibility/resistance to colicin V is not linked to any of the selected markers which have to be given by the F^+ parent.

Many recombinants, however, receive the K resistance characterizing the F^+ parent, because the K marker is linked to the TL markers which must be inherited from the F^+ parent, owing to the selective technique used Still more recombinants are resistant to colicin E, because the E marker is very closely linked to the B_1^+ marker of the F^+ parent (Fredericq & Betz-Bareau, 1952). If, now, selection is made on a minimal medium supplemented with thiamine (i.e. where B_1^+ is no more selected), the number of recombinants which receive the resistance to colicin E from the F^+ parent will fall considerably, but in cases where they do, they will also display the B_1 independence of that parent (Jenkin & Rowley, 1955).

Resistance to a given colicin may sometimes be determined at more than one locus. An auxotrophic mutant derived from the strain *E. coli* B after repeated ultra-violet irradiations (De Haan, 1954) was found to be spontaneously resistant to colicin E. Crosses of this B mutant with a K-12 derivative, also resistant to colicin E, give a proportion of fully susceptible recombinants, probably by recombination of $+ - $ by $- +$ loci.

The occurrence of receptors common to a phage and a given colicin is also confirmed in the recombination experiments. In all crosses studied, there is indeed a perfect correlation between resistance/susceptibility to phage T6 and to colicin K, and between resistance/susceptibility to phage BF23 and to colicin E. In the cross just mentioned between *E. coli* B and K-12, both loci of resistance to colicin E induce resistance to phage BF23 as well, but some recombinants are susceptible to both agents.

Fixation of a colicin on its specific receptor is not sufficient by itself to bring about inhibition. Indeed, some strains may adsorb a given colicin and yet be insensitive to its further action. Susceptible cells, however, are rapidly killed by the colicin they have adsorbed, though the mechanism of this bactericidal action is as yet completely unknown. It is not followed by lysis, although some late and partial lysis, probably due to autolysis of the killed bacteria, may sometimes be observed.

The kinetics of the bactericidal action have been studied with colicins E, K and ML (Fredericq, 1952c; Fredericq & Delcour, 1953; Jacob, Siminovitch & Wollman, 1952, 1953). If the proportion of surviving bacteria is plotted on a log scale against time of action of the colicin, the curve is at first linear but rapidly bends in a plateau, because of the disappearance of the colicin by fixation. The proportion of surviving bacteria too is an exponential function of the colicin concentration, but stands in a direct relation to the number of cells on which the colicin is acting. Survival curves therefore express the adsorption phenomenon underlying colicin action.

Addition of colicin ML to a growing culture of susceptible bacteria greatly disturbs their metabolism (Jacob *et al.* 1952, 1953). Growth ceases at once and RNA and DNA syntheses are immediately stopped. Oxygen uptake remains at its initial rate for about 20 min., then gradually decreases. In fact, colicin seems to block all bacterial syntheses. The addition of colicin to bacteria infected with bacteriophage may also stop the development of the phage, but there sometimes arise phenomena of interference and mutual exclusion between colicins and bacteriophages, comparable to those observed between two phages (Fredericq, 1953 c).

The bactericidal action of colicins is consequently very similar to the action of virulent bacteriophages, except that virulent bacteriophages do not definitely block the syntheses which serve their own reproduction (Jacob et al. 1953). This particularity results from the fundamentally different nature of colicins and bacteriophages. Colicins are inert chemical substances which kill the cells without being reproduced, whereas bacteriophages are biological entities which are reproduced and multiplied by the cells they kill (Fredericq, 1950).

Colicins are perhaps still more particularly related to the lethal protein of these bacteriophages. Virulent bacteriophages have at the tip of their tail a protein component which kills susceptible cells even under conditions where the phage particle does not multiply (Herriott, 1951; Anderson, 1953). Indeed the lethal protein of phage T6, which has the same specificity as colicin K and attaches on the same receptor (Fredericq, 1952 c, d), has also the same susceptibility to X-ray inactivation (Latarjet & Fredericq, 1955) and must, therefore, be a substance related to colicin K. The lethal protein of T6 and colicin K are, however, serologically distinct (Fredericq, 1952 e; Goebel et al. 1955), just as T6 is distinct from other phages having the same specificity of action (Fredericq, 1952 e).

II. COLICINOGENIC FACTORS

The colicinogenic properties are extremely stable hereditary characteristics. The V coli, isolated by Gratia in 1925, still produces the same colicin more than thirty years later, and many strains, studied for over ten years, did not show any variation in their colicinogenic properties. Strains lyophilized and preserved for over fifteen years were found by Nozaki, Robbins & Parr (1953) to produce colicin in the first broth passage.

The colicinogenic properties are independent of the other properties of the strains. For example, many others, biochemical mutants and mutants resistant to phages or to antibiotics, derived from colicinogenic strains, always produced the same colicin as the mother-strain. In one case, however, a lactose-positive mutant derived from a colicinogenic paracoli was found to have lost its colicinogenic property and to be furthermore susceptible to the colicin produced by the mother-strain. The loss of a colicinogenic property is therefore possible, but seems to be quite exceptional. More frequently colicinogenic strains may show quantitative variations in the amount of colicin they produce (Fredericq, 1948e).

In view of their hereditary stability and of their constancy, colicinogenic properties must be governed by genetic determinants, the *colicinogenic factors*, which induce colicin synthesis and ensure its genetic continuity.

Transmissibility of colicinogenic factors

The colicinogenic factors can be transmitted from colicinogenic strains to non-colicinogenic strains in mixed cultures and the transfer may be demonstrated by the following technique: A colicinogenic strain, sensitive to streptomycin, and a non-colicinogenic strain, resistant to this antibiotic, are inoculated in the same broth. After 24 hr. growth, successive dilutions of this mixed culture are spread on to the surface of agar plates containing streptomycin; then a second layer of agar is poured on top. The colonies which appear between the two layers of agar all come from the non-colicinogenic strain, as the other one is inhibited by streptomycin. Those of the colonies which have become colicinogenic are then located by seeding the whole surface of the upper layer, 48 hr. later, with an indicator strain sensitive to colicins (but, of course, resistant to streptomycin). The next day, the confluent culture of this indicator reveals circular inhibition zones centred in the depth by the colicinogenic colonies, which are easy to detect (Fredericq, 1954a).

It is possible by this technique to demonstrate the specific transfer to a strain of non-colicinogenic *E. coli* of the capacity to produce different colicins, using as the initially colicinogenic strain not only *E. coli* but also *Shigella sonnei* (Fredericq, 1954a). The transformed *Escherichia coli* keeps all the properties which characterized the initially non-colicinogenic strain and differs from it only in the newly acquired colicinogenic property. For example, a strain of *E. coli* transformed by mixed culture with a colicinogenic strain of *Shigella sonnei*, retains the features of sensitivity or resistance to phages and to colicins which mark the original strain as well as all properties which distinguish fundamentally the *Escherichia coli* and *Shigella sonnei* species, for example, motility, gas production by glucose fermentation, lactose and xylose fermentation and indol production. In the same way, other enteric bacteria such as *Sh. sonnei* (Fredericq, 1954a), and also paracoli, *K. pneumoniae*, *S. typhi* and *S. para* B (Hamon, 1956b) can be made colicinogenic by mixed culture with a colicinogenic *Escherichia coli* strain.

This isolated genetic transfer of a single trait by mixed culture of two strains comes within the range of phenomena which Lederberg (1952) called transduction. This term is now mostly used in the restricted sense of a genetic transfer mediated by phage particles. As phages do not seem to take any part in the transfer of colicinogenic properties (Fredericq, 1954 b) and to avoid ambiguity, it would perhaps be more appropriate not to use the term transduction to designate the colicinogenic transfer.

The colicinogenic transfer is specific for each colicin considered, and transformed strains always produce a colicin identical to that of the transforming strain (Fredericq, 1954 c; Hamon, 1957 b). For example, the non-colicinogenic strain K-12, transformed by mixed culture with strain CA18, will produce like strains CA18 colicin B. The same K-12 strain, transformed by strain K-235, will produce like K-235 colicin K, and so on.

A colicinogenic property can be serially transferred, for a strain already transformed may in turn transfer its newly acquired colicinogenic property to another strain. A strain already made colicinogenic by a first transformation can still be transformed by a strain producing another colicin and it will then yield two distinct colicins (Fredericq, 1954 c).

The efficiency of transfer varies a great deal according to the case. As a rule, transformed cells represent 1 to 10 % of the originally non-colicinogenic cells. The proportion may reach 100 % when the strain to be transformed is very susceptible to the colicin of the transforming strain. In that case, the non-colicinogenic cells are killed by the colicin, and the only cells to develop in the mixed culture, together with the initially colicinogenic ones, are the transformed cells, which are henceforth immune, as we shall see later, and possibly spontaneous resistant mutants.

Not all colicinogenic strains are able to transfer their colicinogenic properties, and some non-colicinogenic strains cannot be transformed. Among 314 colicinogenic strains tested, only fifty-eight were able to transform non-colicinogenic strains. Among them, thirty-six were doubly colicinogenic but only two could transfer both properties. The ability to transfer seems to depend on the type of colicin, for it can very often be observed with strains producing colicin I, less frequently with strains producing colicins B, E or K, and quite rarely with strains producing colicin V (Fredericq, 1956 a).

The transfer of colicinogenic properties is extremely rapid, at least during the initial step, and can often be demonstrated just by mixing the two strains for a few minutes before diluting and plating.

Although colicinogenic factors are transmissible, they are not released in the medium. The transfer seems to be bound to the presence of living colicinogenic cells and could not be obtained with killed colicinogenic cultures or filtrates, nor even with supernatants of living cultures.

The colicinogenic transfer seems to require conjugation of the cells. Indeed, serial transfers, studied in *E. coli* K-12 derivatives, revealed that the transforming capacity is bound to the F^+ sexual polarity. An F^- strain is not able to transfer its colicinogenic potency unless it is first transformed into F^+ (Fredericq, 1954 *c*).

Independence of the colicinogenic factors from the normal genetic structure of bacteria

Although the colicinogenic transfer requires contact and probably fusion of two cells, it seems to be quite independent of the recombination of the other genetic characters.

Transfer of colicinogenic properties to the fertile *E. coli* K-12 strain made it possible to study the behaviour of the colicinogenic factors in crossing experiments. Most of the parents investigated were derivatives of strain K-12, transformed by the colicinogenic strain K-30, and producing colicin E(1), also known as colicin ER.

Recombination was studied by crossing two complementary auxotrophic parents carrying many differential markers (biochemical properties, resistance to phages, to colicins and to other antibiotics), and by selecting prototrophic recombinants on a synthetic minimal medium (Lederberg's technique, 1947). Recombination in *E. coli* obeys very definite laws (cf. Lederberg, 1955). All genetic markers appear to be linked linearly on a structure comparable to a chromosome, and the frequency of transfer is therefore controlled by their location in relation to the selected markers. The frequency of transfer is also dependent on the F-polarity of the parents, as recombinants result from a crossing over between the whole chromosome of the F^- parent and only a small part of the chromosome of the F^+ parent.

In the crosses, alternatively one or the other parent can be made colicinogenic (Fredericq & Betz-Bareau, 1953 *a*, *b*). Let us take for example the classical cross between strains 58.161 F^+ and W1177 F^-. Strain 58.161 is an auxotrophic mutant, derived from K-12 and requiring methionine but not threonine, leucine and thiamine ($M^-TLB_1^+$). Strain W1177 is the complementary auxotroph requiring threonine, leucine and thiamine but not methionine ($M^+TLB_1^-$). In crosses where the F^- W1177 parent is colicinogenic, all recombinants, without exception, are likewise colicinogenic. Inversion of the colicinogenic property does not, however, reverse its frequency of transfer and many recombinants, up to 60–70 %, are still colicinogenic in the cross where the colicinogenic property is carried by the 58.161 F^+ parent. This is quite unexpected, as a normal marker should have the same frequency of transfer from F^+ to F^-, whether present in the positive or the negative allelic forms.

Even in crosses where the F-polarity is reversed, that is in a cross of an F^- derivative of 58.161 with W1177 which has been made F^+, still 100 % of the recombinants receive the colicinogenic property when it is carried by the F^- parent and around 60 % when it is carried by the F^+ parent.

The same aberrant ratios are also observed in crosses where deficiencies other than methionine or threonine-leucine-thiamine are used as selected markers. Furthermore, the colicinogenic property, which is transmitted from the F^+ to the F^- with a very high frequency, does not appear to be linked to other markers which are also transmitted with such a high frequency. In fact, it is not linked with any of the numerous other markers which were studied.

These results cannot be explained by a lethal effect of the colicinogenic factors, at least in the case of colicin E(1), as these crosses yield about the same number of recombinants as similar crosses where none of the parents is colicinogenic.

In the case of colicinogenic factors governing the production of colicins other than E(1), there is a lethal effect which is more or less pronounced according to the type of colicin, and far fewer recombinants are formed than in reference crosses (Fredericq & Betz-Bareau, 1956). The reduction is already visible in crosses where the F^- parent is colicinogenic, but is most evident when the colicinogenic factor is carried by the F^+ parent. This apparent reduction in fertility is probably due to a colicinogenic induction, similar to the lysogenic induction (Fredericq, 1954e: Jacob & Wollman, 1954) which makes part of the recombinants non-viable. Despite this interference in the apparent fertility, the colicinogenic factors do not, however, influence the frequency of transfer of the other markers.

Some crosses were also studied in which both parents were colicinogenic but for different colicins. The results are essentially the same as in the previous experiments. All recombinants receive the colicinogenic property of the F^- parent. Some of them also receive the colicinogenic property of the F^+ parent and are doubly colicinogenic.

It may be concluded from the aberrant ratios of transfer and from the absence of linkage with any other marker that colicinogenic factors are independent of the normal genetic structure of the bacteria.

Pathogenicity of the colicinogenic factors

Colicinogenic strains are not susceptible to the particular colicin they produce but may, of course, be susceptible to other colicins. This immunity appears to be a direct consequence of the colicinogenic property as non-colicinogenic mutants are susceptible to the colicin produced by the mother-strain (Fredericq, 1948 e).

This immunity applies only to the culture considered as a whole. Colicin is not produced by every cell of a colicinogenic culture. Only cells which do not yield colicin are immune; the others are killed by the colicin they synthesize. They are, however, too few to be manifested and their disappearance is balanced by the development of the first, so that the culture as a whole appears normal. Nevertheless colicin synthesis can be induced in nearly all cells of a colicinogenic culture through the action of many mutagenic or carcinogenic agents, such as ultra-violet rays, peroxides, ethyleneimines or halogenoalkylamines (Jacob *et al.* 1951, 1952, 1953; Lwoff & Jacob, 1952). Colicin induction was first observed in the ML strain of *E. coli* and led to massive lysis of the cells. It was later found that this ML strain is at the same time colicinogenic and lysogenic and that lysis must rather be attributed to development of its prophage (Fredericq, 1954 d; Kellenberger & Kellenberger, 1956). Many other colicinogenic strains, which were obviously non-lysogenic, may be induced to produce colicin but without lysing (Fredericq, 1954 d, 1955; Hamon & Lewe, 1955; Kellenberger & Kellenberger, 1956; Panijel & Huppert, 1956). Colicins indeed are bactericidal but not bacteriolytic agents.

The induction of colicin synthesis in colicinogenic strains is very similar to the induction of prophage development in lysogenic strains (Jacob *et al.* 1953). In both cases the induction kills the cells after a period of about 90 min., during which there is a residual growth with a parallel increase in respiratory intensity. Colicin liberation starts, however, earlier, as phage production is only manifested after lysis of the cells (Fredericq, 1954 d). Other syntheses continue at a reduced rate, except that of DNA, which is temporarily blocked in lysogenic induction. This difference is probably due to the fact that bacteriophages contain a DNA fraction, whereas colicins are basically of a protein nature.

Like lysogenic factors, colicinogenic factors are therefore potential lethal agents whose pathogenicity is only disclosed by the achievement of their potentiality. As long as they remain in a latent condition, they induce, on the contrary, immunity to the corresponding colicin.

The same type of potential pathogenicity and immunity is also verified in originally non-colicinogenic strains which are made colicinogenic by mixed culture with a colicinogenic one (Fredericq, 1954 d; Hamon, 1957 a). The introduction of a colicinogenic factor into transformed cells brings along immunity to the colicin whose synthesis it controls, but results in the killing of the cells where that synthesis succeeds.

This specifically acquired resistance of transformed cells to the colicin they now produce is not achieved through the loss of the corresponding receptor, for transformed cells keep the receptor to that colicin if they had it already before transformation. For example, when a strain possessing the receptor for colicins of group E is transformed by a strain producing a colicin of that group, it is no longer receptive to that particular colicin. It keeps, however, the receptor, for it remains susceptible to the other colicins of the group, as well as to phage BF23, which are adsorbed by the same receptor. On the other hand, when resistance is achieved by mutation through loss of the receptor, the resistant mutants always resist all colicins of the group as well as phage BF23. Immunity is also more specific than resistance and may serve to discriminate colicins which are adsorbed by the same receptor (Fredericq, 1956b).

Immunity and resistance may be further differentiated by the fact that they behave as non-allelic markers in recombination experiments. Crosses of two parents which are both non-receptive to the same colicin but for different reasons, the one because it produces that colicin and the other because it has lost the corresponding receptor, may yield cells which recombine presence of the receptor, received from the immune parent, and absence of immunity, inherited from the resistant one, and are therefore fully susceptible (Fredericq, unpublished).

Immunity of colicinogenic strains appears to be identical to that of lysogenic strains. In both cases, it results from the presence of a potential lethal agent, but persists only as long as this agent remains in a latent condition. It is perhaps easier to conceive of the immunity of lysogenic, than of colicinogenic bacteria. In lysogenic bacteria, the presence of a DNA structure, the prophage, prevents the development of related DNA structures introduced into the cells. A single mechanism is probably responsible for hindering the development of the carried prophage as well as of the related infecting particles. In colicinogenic bacteria, on the other hand, the presence of the colicinogenic factor, does not prevent the development of similar structures but the action of a quite different agent, the colicin whose synthesis it potentially controls. This is an unprecedented example of cellular immunity to a well-defined chemical substance, but nothing is yet known about its mechanism.

Immunity of colicinogenic strains to their particular colicin is rarely complete and depends on which type of colicin is produced. Strains made colicinogenic for a colicin of group E are immune to that colicin, even if they keep the corresponding receptor. They tend, however, to lose it in the course of subcultures, probably because cells that keep it are somewhat inhibited. Strains producing colicin I are resistant to the level of colicin they release but may be inhibited by the same colicin at higher concentration. Immunity to colicin V is still less pronounced and strains made V-colicinogenic are partially inhibited by their own colicin. Their broth cultures are less turbid than those of the same strains before transformation; agar streaks develop poorly and isolated colonies are small and irregular (Fredericq, 1956 b). A strain spontaneously colicinogenic and susceptible to the colicin it produces has, moreover, been described (Ryan, Fried & Mukai, 1955).

III. DISCUSSION AND CONCLUSION

Many coli and related bacteria produce a variety of diffusible antibiotic substances, named colicins, which are all of a protein or polypeptide nature but differ by many other characteristics. Their activity-spectra are strikingly different although strictly limited to other strains of the family Enterobacteriaceae. They kill susceptible bacteria after fixation on specific receptors.

Colicins display unexpected relations with bacteriophages. Activity-spectra of both agents are quite comparable in their diversity, and susceptibility-patterns are controlled by a number of specific receptors. Some of these receptors are even common to a phage and a given colicin. The bactericidal action of colicins is comparable to that of virulent bacteriophages. In fact, some virulent bacteriophages have a bactericidal component which is a protein related to colicins.

Colicins and bacteriophages are, however, agents of a fundamentally different nature. Colicins are inert chemical substances, proteins, which kill bacteria without being reproduced, whereas bacteriophages are biological units, endowed with genetic continuity, which are reproduced and multiplied by the cells they kill. Phages are much more complex particles which include a deoxyribonucleic core and a protein envelope.

The genetic structures, which are included in the phage particles and induce their reproduction, are absent from colicins. Owing to the hereditary stability and specificity of the colicinogenic properties, such genetic structures must, however, be present in the producing bacteria. These *colicinogenic factors* may in fact be transmitted from one strain to another but, as they are not set free by lysis of the cells, their transfer requires the mechanism of sexual conjugation. Yet they behave in recombination experiments as if completely independent of the normal genetic structure of the bacteria.

Like prophages, colicinogenic factors are potential lethal agents whose spontaneous or induced development kills the cells in which it succeeds. They are also pathogenic, although their pathogenicity is only disclosed by the achievement of their potentiality. As long as they remain in a latent condition they induce, on the other hand, immunity to the corresponding colicin.

Colicinogenic factors are transmissible pathogenic agents which are independent of the normal genetic structure of bacteria and could therefore be regarded as bacterial viruses, distinct from bacteriophages. Their relations with the latter point perhaps to a common parental ancestry.

REFERENCES

ABBOTT, J. D. & SHANNON, R. (1956). *J. Path. Bact.* **72**, 350.

ANDERSON, T. F. (1953). *Cold Spr. Harb. Symp. Quant. Biol.* **18**, 197.

BLACKFORD, V. L., PARR, L. W. & ROBBINS, M. L. (1951). *Antibiot. Chemother.* **1**, 392.

BORDET, P. (1947). *Rev. Immunol.* **2**, 323.

BORDET, P. (1948). *C.R. Soc. Biol., Paris,* **142**, 257.

BORDET, P. & BEUMER, J. (1948). *C.R. Soc. Biol., Paris,* **142**, 259.

BORDET, P. & BEUMER, J. (1951). *Rev. Belge Path.* **21**, 245.

CHABBERT, Y. (1950). *Ann. Inst. Pasteur,* **79**, 51.

COOK, M. K., BLACKFORD, V. L., ROBBINS, M. L. & PARR, L. W. (1953). *Antibiot. Chemother.* **3**, 195.

DE HAAN, P. G. (1954). *Genetica,* **27**, 300.

DEPOUX, R. & CHABBERT, Y. (1953). *Ann. Inst. Pasteur,* **84**, 798.

FASTIER, L. B. (1949). *J. Immunol.* **62**, 399.

FREDERICQ, P. (1946a). *C.R. Soc. Biol., Paris,* **140**, 1033.

FREDERICQ, P. (1946b). *C.R. Soc. Biol., Paris,* **140**, 1189.

FREDERICQ, P. (1946c). *Schweiz. Z. Path.* **9**, 385.

FREDERICQ, P. (1948a). *C.R. Soc. Biol., Paris,* **142**, 399.

FREDERICQ, P. (1948b). *C.R. Soc. Biol., Paris,* **142**, 403.

FREDERICQ, P. (1948c). *C.R. Soc. Biol., Paris,* **142**, 853.

FREDERICQ, P. (1948d). *C.R. Soc. Biol., Paris,* **142**, 855.

FREDERICQ, P. (1948e). *Rev. Belge Path.* **19**, suppl. 4, 1.

FREDERICQ, P. (1949). *C.R. Soc. Biol., Paris,* **143**, 1011.

FREDERICQ, P. (1950). *Bull. Acad. Méd. Belg.* **15**, 491.

FREDERICQ, P. (1951). *Leeuwenhoek ned. Tijdschr.* **17**, 227.

FREDERICQ, P. (1952a). *C.R. Soc. Biol., Paris,* **146**, 298.

FREDERICQ, P. (1952b). *Rev. Belge Path.* **22**, 167.

FREDERICQ, P. (1952c). *C.R. Soc. Biol., Paris,* **146**, 1295.

FREDERICQ, P. (1952d). *C.R. Soc. Biol., Paris,* **146**, 1406.

FREDERICQ, P. (1952e). *C.R. Soc. Biol., Paris,* **146**, 1624.

FREDERICQ, P. (1953a). *Ann. Inst. Pasteur,* **84**, 294.

FREDERICQ, P. (1953b). *Bull. Acad. Méd. Belg.* **18**, 126.

FREDERICQ, P. (1953c). *C.R. Soc. Biol., Paris,* **147**, 533.

FREDERICQ, P. (1954a). *C.R. Soc. Biol., Paris,* **148**, 399.

FREDERICQ, P. (1954b). *C.R. Soc. Biol., Paris,* **148**, 624.

FREDERICQ, P. (1954c). *C.R. Soc. Biol., Paris,* **148**, 746.

FREDERICQ, P. (1954d). *C.R. Soc. Biol., Paris,* **148**, 1276.

FREDERICQ, P. (1954e). *C.R. Soc. Biol., Paris,* **148**, 1501.

FREDERICQ, P. (1955). *C.R. Soc. Biol., Paris,* **149**, 2028.

FREDERICQ, P. (1956a). *C.R. Soc. Biol., Paris,* **150**, 1036.

FREDERICQ, P. (1956b). *C.R. Soc. Biol., Paris*, **150**, 1514.
FREDERICQ, P. & BETZ-BAREAU, M. (1948). *C.R. Soc. Biol., Paris*, **142**, 1180.
FREDERICQ, P. & BETZ-BAREAU, M. (1952). *Ann. Inst. Pasteur*, **83**, 283.
FREDERICQ, P. & BETZ-BAREAU, M. (1953a). *C.R. Soc. Biol., Paris*, **147**, 1653.
FREDERICQ, P. & BETZ-BAREAU, M. (1953b). *C.R. Soc. Biol., Paris*, **147**, 2043.
FREDERICQ, P. & BETZ-BAREAU, M. (1956). *C.R. Soc. Biol., Paris*, **150**, 615.
FREDERICQ, P., BETZ-BAREAU, M. & NICOLLE, P. (1956). *C.R. Soc. Biol., Paris*, **150**, 2039.
FREDERICQ, P. & DELCOUR, G. (1953). *C.R. Soc. Biol., Paris*, **147**, 1310.
FREDERICQ, P. & GRATIA, A. (1950). *Leeuwenhoek ned. Tijdschr.* **16**, 119.
FREDERICQ, P., JOIRIS, E., BETZ-BAREAU, M. & GRATIA, A. (1949). *C.R. Soc. Biol., Paris*, **143**, 556.
FREDERICQ, P. & LEVINE, M. (1947). *J. Bact.* **54**, 785.
GARDNER, J. F. (1950). *Brit. J. Exp. Path.* **31**, 102.
GOEBEL, W. F., BARRY, G. T., JESAITIS, M. A. & MILLER, E. M. (1955). *Nature, Lond.* **176**, 700.
GOEBEL, W. F., BARRY, G. T. & SHEDLOVSKY, T. (1956). *J. Exp. Med.* **103**, 577.
GRATIA, A. (1925). *C.R. Soc. Biol., Paris*, **93**, 1040.
GRATIA, A. (1932). *Ann. Inst. Pasteur*, **48**, 113.
GRATIA, A., FREDERICQ, P., JOIRIS, E., BETZ-BAREAU, M. & WEERTS, E. (1950). *Leeuwenhoek ned. Tijdschr.* **16**, 31.
GROSSO, E. (1950). *Boll. Ist. Sieroter. Milano*, **29**, 373.
GUELIN, A. (1943). *Ann. Inst. Pasteur*, **69**, 382.
HALBERT, S. P. (1948a). *J. Immunol.* **58**, 153.
HALBERT, S. P. (1948b). *J. Immunol.* **60**, 23.
HALBERT, S. P. (1948c). *J. Immunol.* **60**, 359.
HALBERT, S. P. & GRAVATT, M. (1949a). *J. Immunol.* **61**, 271.
HALBERT, S. P. & GRAVATT, M. (1949b). *Publ. Hlth Rep., Wash.* **64**, 313.
HALBERT, S. P. & MAGNUSON, H. J. (1948). *J. Immunol.* **58**, 397.
HAMON, Y. (1955). *Ann. Inst. Pasteur*, **88**, 193.
HAMON, Y. (1956a). *C.R. Acad. Sci., Paris*, **242**, 1240.
HAMON, Y. (1956b). *C.R. Acad. Sci., Paris*, **242**, 2064.
HAMON, Y. (1957a). *Ann. Inst. Pasteur*, **92**, 363.
HAMON, Y. (1957b). *Ann. Inst. Pasteur*, **92**, 489.
HAMON, Y. & LEWE, Z. (1955). *Ann. Inst. Pasteur*, **89**, 336.
HEATLEY, N. G. & FLOREY, H. W. (1946). *Brit. J. Exp. Path.* **27**, 378.
HERRIOTT, R. M. (1951). *J. Bact.* **61**, 752.
JACOB, F., SIMINOVITCH, L. & WOLLMAN, E. (1951). *C.R. Acad. Sci., Paris*, **233**, 1500.
JACOB, F., SIMINOVITCH, L. & WOLLMAN, E. (1952). *Ann. Inst. Pasteur*, **83**, 295.
JACOB, F., SIMINOVITCH, L. & WOLLMAN, E. (1953). *Ann. Inst. Pasteur*, **84**, 313.
JACOB, F. & WOLLMAN, E. (1954). *C.R. Acad. Sci., Paris*, **239**, 317.
JENKIN, C. R. & ROWLEY, D. (1955). *Nature, Lond.* **175**, 779.
KELLENBERGER, G. & KELLENBERGER, E. (1956). *Schweiz. Z. Path.* **19**, 582.
LATARJET, R. & FREDERICQ, P. (1955). *Virology*, **1**, 100.
LEDERBERG, J. (1947). *Genetics*, **32**, 505.
LEDERBERG, J. (1952). *Physiol. Rev.* **32**, 403.
LEDERBERG, J. (1955). *J. Cell. Comp. Physiol.* **45**, 75.
LEVINE, M. & TANIMOTO, R. H. (1954). *J. Bact.* **67**, 537.
LUDFORD, C. G. & LEDERER, M. (1953). *Aust. J. Exp. Biol. Med. Sci.* **31**, 553.
LWOFF, A. & JACOB, F. (1952). *C.R. Acad. Sci., Paris*, **234**, 2308.
MONDOLFO, U. (1948). *Boll. Soc. Ital. Biol. Sper.* **24**, 1.
MONDOLFO, U. (1949). *Anales Med. Publica*, **1**, 219.

MONDOLFO, U. & CEPPELLINI, R. (1950). *Boll. Ist. Sierot. Milano*, **29**, 231.

NOZAKI, J. N., ROBBINS, M. L. & PARR, L. W. (1953). *J. Bact.* **66**, 621.

PANIJEL, J. & HUPPERT, J. (1956). *C.R. Acad. Sci., Paris*, **242**, 199.

PARMALA, M. E. (1956). *Ann. Med. Exp. Fenn.* **34**, suppl. 7, 1.

RETTGER, L. F. (1928). In *The Newer Knowledge of Bacteriology and Immunology*. By Jordan, E. O. & Falk, I. S. Chicago (Univ. of Chicago Press).

ROBBINS, M. L., EL SHAWI, N. N. & PARR, L. W. (1956). *Bact. Proc.* **97**, 97.

RYAN, F. J., FRIED, P. & MUKAI, F. (1955). *Biochim. Biophys. Acta*, **18**, 131.

ZAMENHOF, S. (1945). *J. Bact.* **49**, 413.

REPLICATION OF AN ANIMAL VIRUS

By F. K. SANDERS, J. HUPPERT* AND J. M. HOSKINS

Medical Research Council Virus Research Group, London
School of Hygiene and Tropical Medicine

One fundamental feature of a self-replicating system is that it can only be made from others of the same kind (Pontecorvo, this Symposium). Although we can imagine self-reproducing systems which consist of chains or cycles of reactions which do not need to be organized in space (Hinshelwood, 1947, 1952; Delbruck, 1949; Pollock, 1953), most biologists familiar with the structural complexity of cells cannot readily accept such growing bags of enzymes as the commonest form of self-replicating system. A more intellectually satisfying model includes the additional idea of replication of a specific structural pattern, this pattern being responsible not only for its own replication but also for that of other patterns which in some way depend upon it. In this connexion the model for the structure of deoxyribonucleic acid (DNA) given by Watson & Crick (1953) was of great importance because it provided for the first time ideas about how a self-duplicating pattern of this kind could function in its own duplication. And now Crick (this Symposium) has further extended these ideas to show how a nucleic-acid pattern could also determine the manufacture of dependent patterns—i.e. proteins. One way to investigate this second type of self-replication in detail is to introduce alien self-duplicating patterns into cells, which are then produced by the cells together with their dependent patterns. An opportunity for studying problems of this kind is given by the interactions between viruses and their host cells. In fact, much of the support for current ideas about the behaviour of self-replicating systems comes from intensive studies of bacteriophage, in particular the bacteriophage T2 (Hershey, 1956).

The nucleic acid of this phage is entirely DNA. It is, moreover, a large phage with a complex morphology reflected both in its appearance under the electron microscope (Williams, 1957), and in its antigenic structure (Lanni & Lanni, 1953). This point is stressed because it is perhaps premature to assume that all phages will turn out to be similarly constructed, or to behave in exactly the same way within their host cells; especially the smaller phages of the T series (T1, T7) or the extremely minute (15 mμ) *Shigella* phage mentioned by Thomas (this Symposium). Still less is it

* Exchange Scholar, Centre National de la Recherche Scientifique, France. Present address: Institut Pasteur, Paris.

to be expected that viruses infecting animal cells, which lack the complex head-and-tail differentiation of T2 bacteriophage, and which contain ribonucleic acid (RNA) rather than DNA, will mirror the reproductive behaviour of T2.

One of the most seductive aspects of the *Escherichia coli* B-T2 bacteriophage system is the ease with which quantitative studies can be made. A known number of susceptible cells, believed to be in a similar physiological state, can be exposed simultaneously to a known number of virus particles, and the resulting events followed quantitatively. It is much more difficult to make comparable studies with viruses which grow intracellularly in higher organisms. The widespread use of tissue culture methods during the last five years (see review by Dulbecco, 1955), especially the development of 'plaque' methods for the assay of animal viruses (Dulbecco & Vogt, 1954 *a*, *b*), has provided many new observations concerning the mode of interaction of these agents with their host cells. However, many of these observations still lack the precision of those obtainable with bacterium-bacteriophage systems. This is chiefly due to the fact that most studies of animal viruses in tissue cultures depend on sheets of cells attached to glass surfaces. The constituent cells of such monolayers, prepared by the tryptic digestion of chick embryos or of whole organs such as the kidney, or other cell cultures, are often not all of the same kind or in the same physiological condition. Although studies have been made using both populations of separated cells (Dulbecco & Vogt, 1954 *a*; Pereira, 1954) and single cells in isolation (Lwoff, Dulbecco, Vogt & Lwoff, 1955), we do not know to what extent the treatment used in preparing the cells may influence their eventual response to infection. For this reason it was thought worth describing some studies our group have made during the past year with a new animal virus-host cell system which comes close to the bacteriophage-bacterium system with regard to the ease with which both cells and virus can be enumerated. The virus used was the virus of encephalomyocarditis (EMC). This is a small (25–30 mμ) virus, probably containing only RNA and protein, and is a natural parasite of rodents; in infected mice it causes a severe central nervous system disease which is invariably fatal. The cells used were those of the Krebs 2 ascites carcinoma of mice (Klein & Klein, 1951) maintained *in vitro*. Mice inoculated intraperitoneally with tumour cells produce up to 15 ml. of an exudate consisting of a suspension of 1–2 × 10^8/ml. of free, separate, cells, over 95 % of which are viable. By centrifuging at low speed the cells can easily be obtained free from both the ascitic fluid and the small amount of erythrocytes which the exudate normally contains.

CHARACTERISTICS OF THE ENCEPHALOMYOCARDITIS (EMC) VIRUS KREBS 2 CELL SYSTEM

Hoskins & Sanders (1957) have shown that Krebs 2 ascites cells in stationary tube cultures become attached to the glass, and are able to support multiplication of EMC virus with the production of typical cytopathic change. The maximum cell population which can survive such culture conditions is about 10^6 cells/ml. These tumour cells can, however, also be maintained as a suspension of separate cells in continuously swirled cultures; under such conditions the cells will survive at concentrations of up to 2×10^7 cells/ml. for at least 48 hr., although they do not appear to grow. EMC virus can multiply

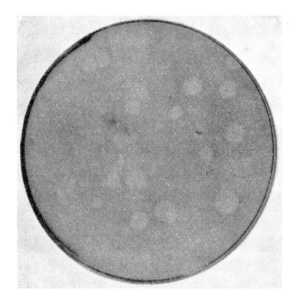

Fig. 1. Plaques of EMC virus (K2 strain) on an agar layer of Krebs 2 carcinoma cells.

in such cultures with the production of high yields of virus accompanied by cell destruction. Suspended tumour cells can also be used to titrate the infectivity of EMC virus by means of a 'plaque' technique, analogous to that described by Cooper (1955). Aliquots containing nutrients of cell suspensions infected with a few virus particles are mixed with agar and poured into Petri dishes on top of a lower layer of agar. After 2–3 days' incubation, the plates are stained with neutral red (which stains only living cells). The foci of infection then appear as pale areas on a red-stained background of normal cells. Fig. 1 shows typical plaques produced by EMC virus in this way.

In addition to its power of producing progeny as measured by the plaque method, other properties of the virus can be measured. One of these is its power to kill cells independently of its power to multiply within them. This 'cell-killing' property is not a primary toxic effect. Death of the affected cells only takes place after an interval equivalent to one virus growth cycle, the phenomenon being analogous to the cell-killing effect of bacteriophage described by many authors (see Herriott, 1951).

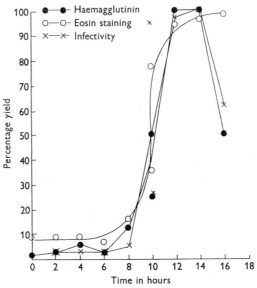

Fig. 2. Correlation between cell staining by eosin and the release of new virus in suspensions of Krebs 2 carcinoma cells infected with EMC virus.

The cell-killing effect of EMC virus can be detected in a simple way. Schrek (1936) showed that when suspensions of tumour cells are stained briefly with 0·05 % eosin, only non-viable cells stain. Hoskins, Meynell & Sanders (1956) have shown a similar correlation between eosin staining and cell viability in the case of Krebs 2 carcinoma cells. In swirling cultures of Krebs 2 cells up to 36 hr. old, less than 5 % of cells stain. However, when such cultures are infected with EMC virus a wave of 'stainability' passes through the culture about 10 hr. after infection, which is correlated with the appearance of new virus in the extracellular fluid. This is shown in Fig. 2. Eosin staining can thus be used to mark those cells in an infected culture which have been killed by virus, and Sanders (1957) has described a method for titrating the cell-killing power of EMC virus based on this correlation.

EMC virus in addition has the power of agglutinating mammalian erythrocytes in the cold, and haemagglutinin titrations can be used to estimate the virus, by methods such as those described by Gard & Heller (1951).

To summarize, EMC virus can be grown in agitated cultures of Krebs 2 carcinoma cells where almost all cells remain separate, susceptible to virus action, and able to be counted simply in a haemocytometer. When virus is grown in such agitated cultures the following three properties of the virus can be measured: (a) its power to produce progeny virus (i.e. plaques); (b) its power to kill cells independently of plaque-forming ability; (c) haemagglutinin.

The following sections describe what happens when a population of ascites tumour cells in swirling culture is infected with EMC virus and the course of events followed by means of these techniques.

INITIAL STAGES OF INFECTION

When a population of Krebs 2 cells is infected by EMC virus the first event which can be detected is the appearance within the cell population of increasing numbers of 'infected' cells. Their number at any time after infection can be measured by standard bacteriophage techniques (Adams, 1950), such as the elimination of extracellular virus with antiserum, followed by plating for 'infectious centres'.

Fig. 3 shows the effect of adding virus to 10^7 cells/ml. in the proportion of approximately one virus particle per 10 cells. At this ratio of infection only about 10 % of the cells can become infected, and only a very small proportion of these will become infected by more than one particle; the number of infected cells is thus also a measure of the number of successful virus particles. Under these conditions up to 95 % of the virus is successful in initiating infection in the first 15 min.

Fig. 4 shows the results of a similar experiment in which the 'cell-killing' rather than the 'plaque-forming' power of the virus was measured. A much more concentrated cell suspension (10^8/ml.) was used, and virus added at the outset to a concentration 2–3 times greater than that of the cells. The lowermost curve (Fig. 4, open circles) shows the rate of appearance in the tumour cell culture of those cells which were eventually killed by virus. The middle curve (filled circles) is a measure of the amount of cell-killing property actually 'adsorbed' by the cells; the distance between it and the lower curve is a measure of the 'multiplicity', or the average number of 'cell-killing' particles adsorbed by each cell, in this case between two and three. The uppermost curve (crosses) illustrates the disappearance of cell-killing virus from the extracellular fluid; the points on this curve were obtained from separate titrations of the cell-killing power of timed samples of the basic culture from which the cells had been removed by centrifugation. All three curves of Fig. 4 show that the cell-killing property of a virus suspension, like its plaque-forming ability, becomes rapidly attached to the cells, although perhaps more slowly than the latter. Under the conditions of this experiment virtually all the cell-killing power is so adsorbed.

When this initial phase of infection is followed by measuring the virus haemagglutinin, a somewhat paradoxical result is obtained. Fig. 5 shows

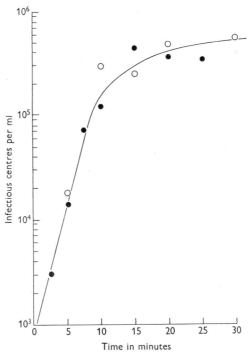

Fig. 3. Adsorption of the plaque-forming ability of EMC virus in infected suspensions of Krebs 2 carcinoma cells. The open and filled circles represent data from two different experiments.

the haemagglutinin content of both the cellular and extracellular fractions of a culture (10^7 cells/ml.) infected with 2–3 plaque-forming particles per cell. It will be seen that although there is clear evidence that the cells have become infected, as shown by the appearance of new haemagglutinin in the cellular fraction, 3–4 hr. later, there appeared to be no diminution in the amount of extracellular haemagglutinin originally present. This result has been abundantly confirmed in other experiments. Therefore, while the cell-killing and plaque-forming ability of the virus appears to become

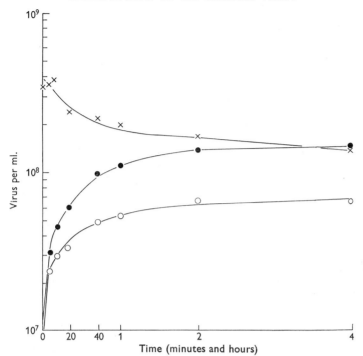

Fig. 4. Adsorption of the 'cell-killing' property of EMC virus in an infected suspension of Krebs 2 cells. Explanation in text.

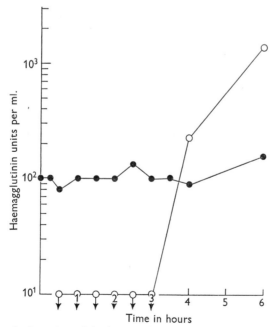

Fig. 5. Failure of adsorption of the haemagglutinin of EMC virus in an infected suspension of K2 cells. Filled circles: 'extracellular' haemagglutinin. Open circles: 'cell-associated' haemagglutinin.

readily attached to the cells at the outset of infection, this does not appear to be the case with virus haemagglutinin.

At least three explanations of this apparent paradox are possible. First of all, it can be suggested that the haemagglutinating and infective virus particles are not identical. Although not excluded, this seems unlikely, since all former work on EMC virus haemagglutination has indicated that both infectivity and haemagglutination are properties of the same kind of particles. For example, these properties sediment together in the ultra-centrifuge and adsorb to red cells in the same way. (Gard & Heller, 1951; Horvath & Jungeblut, 1952). Moreover, antisera against EMC virus both inhibit haemagglutination and neutralize infectivity; with any set of sera there is a parallelism between haemagglutination inhibition and neutralizing titres, which indicates that similar antigens may be involved.

A second, more plausible, hypothesis to account for the anomalous behaviour of the haemagglutinin would be that all virus preparations contain so large an excess of non-infectious haemagglutinin that the small fraction of plaque-forming and cell-killing particles actually attached to the Krebs cells during infection, cannot be detected in haemagglutination tests. This suggestion is more difficult to disprove than the former and is not, as yet, completely ruled out. However, the following considerations suggest that it is not the correct explanation of the effect observed. First, if samples of an infected culture of Krebs 2 cells are chilled, the cells spun down, and haemagglutinin estimated in the supernatant, a considerable amount of adsorption can sometimes be demonstrated. Secondly, preparations of EMC virus treated in many different ways have about the same ratio of cell-killing to haemagglutinating effect (about 5×10^5 cell-killing particles/ haemagglutinating unit). At the end-point of the haemagglutination titration, about 10% of the 10^7 red cells added are agglutinated. If we assume that these 10^6 agglutinated cells consist of aggregates of pairs of cells, each stuck together by a single virus particle, the suspension should contain about $10^6/2$, i.e. 5×10^5, haemagglutinating particles. This calculation suggests that there may not be, in fact, a great excess of haemagglutinating particles in EMC virus suspensions.

A third hypothesis to account for the failure of haemagglutinin adsorption is that infection takes place during a very transient contact of virus and cell. Shortly afterwards the haemagglutinin is released into the extracellular fluid. The hypothesis does not imply that there is any event directly comparable to the injection process observed with bacteriophage T2 (Hershey & Chase, 1952). EMC virus particles may even penetrate into the interior of the cells and the haemagglutinating residue be expelled later on.

During the short-lived contact between cell and virus 'something' passes from the virus to the cell which initiates two processes, one resulting eventually in cell death, and one in the production of new infective virus. It is necessary to make a formal distinction between these two trains of events since there is some evidence that the 'cell-killing' and 'plaque-forming' properties of the virus are affected differently by various experimental procedures. Mistreatment of the virus in various ways—i.e. by heating, freezing to $-40°$ C., and perhaps prolonged mechanical agitation, can reduce the plaque-forming titre of a virus preparation by a factor of at least 10–100 times, while the cell-killing titre falls only twofold; the haemagglutinin is largely unaffected by such procedures. A virus preparation treated in these ways thus contains an excess of particles which can adsorb transiently to cells and which can start a process which kills them, but cannot produce offspring. Other cases of what may be a similar process can be found in the literature. For example, Ackermann, Rabson & Kurtz (1954) working with HeLa cell cultures massively infected with type III poliomyelitis virus, found a latent period of about 4–5 hr., after which release of new virus lasted from 6–7 hr. When virus multiplication was blocked by an analogue of phenylalanine (fluoro-phenylalanine) which stops virus growth at an early stage, the cytopathic effect was found nevertheless to appear at the proper time. Similarly, Henle, Girardi & Henle (1955) studied the multiplication of influenza virus in HeLa cells. Cytopathic effects appeared 12–96 hr. after infection but no infective virus was produced. However, they found that haemagglutinin and complement-fixing antigen were produced inside cells although little was released extracellularly. Poliomyelitis virus does not possess a demonstrable haemagglutinin, and we have not yet looked for the development of haemagglutinin or complement-fixing antigen in Krebs 2 cells treated with virus with a high cell-killing to plaque-forming ratio. In this case also, there may still be an abortive cycle of infection. However, the argument that cytopathic changes are not invariably accompanied by the release of active progeny still stands, and thus the idea that the 'something' which penetrates the cells starts more than one process. Moreover, the fact that the biosynthetic process *starts* at all under these conditions indicates that the virus must have some triggering effect on the cell.

THE NATURE OF THE PENETRATING MATERIAL

One may speculate that the material which penetrates into the cells, and which carries the information corresponding to the cell-killing and plaque-forming properties, is perhaps the virus RNA. At the present time there is no clear evidence that this is so. Gierer & Schramm (1956) were the first to demonstrate that RNA preparations of tobacco mosaic virus are infective and can give rise to virus progeny with the properties of the original strain from which they were prepared. Similarly, Colter, Bird & Brown (1957) have claimed that RNA preparations from Ehrlich ascites carcinoma cells infected with Mengo encephalitis virus are infective for mice, but the experiments quoted in their brief report do not fully establish their claim, and further work is needed. However, Wecker & Schäfer (1957) have recently demonstrated that RNA preparations made from the brains of mice moribund following infection with Eastern Equine Encephalitis may be infective. We have also found that RNA extracts made from Krebs 2 cells infected with EMC virus contain small numbers (less than 10^2/ml.) of plaque-forming units. Infective RNA preparations, however, have only been obtained from preparations which themselves contain large amounts of active virus (10^8 PFU/ml.) and it cannot as yet be ruled out that this 'infective RNA' is not residual virus. More work is needed on this point. Some indirect evidence, however, indicates that there may be a 'naked RNA' phase at the beginning of the growth cycle in the case of some viruses. Hamers-Castermann & Jeener (1957) showed in the case of tobacco mosaic virus that ribonuclease given up to 2 hr. after infection suppressed the eventual development of virus. Their study was carefully controlled to show that the effect observed was due to enzyme action and not to complex formation with the virus. Similarly Le Clerc (1956) also showed that the intracellular multiplication of influenza virus is suppressed by ribonuclease if the latter is given up to 2 hr. after infection.

If RNA is the carrier of the cell-killing and plaque-forming information we can estimate how much is needed per infected cell. Infection of a cell by a single virus particle is enough to initiate production of both plaque-forming and cell-killing material. The absolute amount of RNA in one virus particle is about the same as that for other virus particles and also for microsomes (Frisch-Niggemeyer, 1956). Thus, an amount of information equivalent to that contained in one microsome, 10^{-3}–10^{-4} of the amount already present in the cytoplasm, is enough to initiate the whole process. However, it must be remembered that we have as yet no clear evidence that RNA is the only thing which enters the cell. Even in the case of T2 bacteriophage, which injects its charge of DNA into the susceptible bacteria, Hershey (1956) now stresses the fact that up to 3 % of the phage protein also enters the cells, and we are not yet certain of the part it may play in the subsequent copying process. Perhaps the protein injected may be responsible for the triggering effect mentioned earlier.

LATER EVENTS IN THE CYCLE OF INFECTION

Following infection EMC virus undergoes a true 'eclipse' phase. This means that for a period of up to 3 hr. after invasion of the cells less than 0·1 plaque-forming virus particles per infectious centre can be found associated with the cell fraction of a culture. New virus first appears within the cells

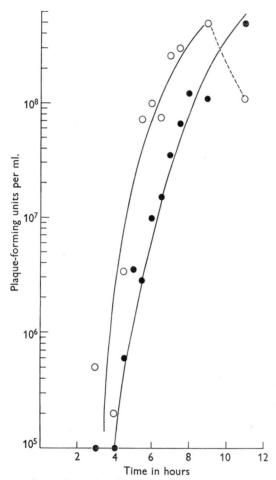

Fig. 6. Appearance of new plaque-forming virus in a suspension of Krebs 2 cells infected with EMC virus. Open circles: 'cell-associated' virus. Filled circles: extracellular' virus.

3–4 hr. after infection and increases at a rapid rate intracellularly. All three properties reappear roughly at the same time. These events are summarized in Fig. 6. The virus concentration within the cells builds up to a high level before significant amounts of new virus are released into the fluid phase of the culture. This aspect of the growth cycle is not clearly shown in Fig. 6, where the virus concentrations are shown plotted on a logarithmic scale. In Fig. 7, where the intra- and extracellular virus concentrations are shown as percentages of the final yield obtained, it can be seen that nearly all of the virus formed is present intracellularly before significant amounts are released from the infected cells. Release of new virus begins at about 5 hr. after infection and is complete by 12 hr. In a large population of infected cells virus is being most rapidly released between 10 and 11 hr. after infection. The figures given in Fig. 7 represent the release of virus from a

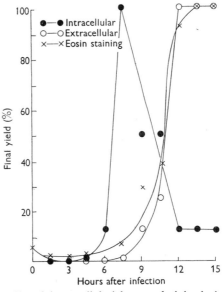

Fig. 7. 'Cell-associated' and 'extracellular' haemagglutinin during the phase of virus maturation and release in Krebs 2 cells infected with EMC virus. Correlation with eosin staining.

whole population of infected cells. Experiments in which the release of virus from very small numbers of infected cells (less than 5) were studied have shown that individual cells release their virus content very rapidly. A few virus particles leak out at first, but are followed shortly afterwards by the rapid release of over 200 new plaque-forming units per cell. At the time of this 'burst' the cells become stainable by eosin and the cytopathic effect appears. All three virus properties appear to be released from the infected cells at the same time. In this respect EMC virus resembles poliomyelitis virus (Lwoff et al. 1955), and is unlike Western Equine Encephalomyelitis (Rubin, Baluda & Hotchin, 1955) or Newcastle disease (Rubin, Franklin & Baluda, 1957). In the case of W.E.E., mature virus is released from the cells as soon as it is formed, so that at any time during the phase of virus maturation each cell does not contain more than 4–10 virus particles. In the case of NDV the mature virus remains associated with the cells for about 80 min. During this period the virus is close to the cell surface, since it can be shown to be sensitive to the action of anti-serum. Electron microscope evidence confirms this fact, and also suggests that viruses of the influenza group may undergo maturation at the cell surface. By contrast, the maturation phase of EMC virus is probably intra-cellular, in view of the high intracellular virus concentration found, and the fact that this phase is not antiserum-sensitive.

GENETIC ASPECTS OF EMC INFECTION

Pontecorvo (this Symposium) gives a second criterion of a self-replicating system—namely that it can give origin to a 'mutated' system, which in its turn is capable of copying itself in the mutated form. This criterion also is fulfilled by the EMC virus-tumour cell system. Starting with EMC virus of mouse-brain origin two strains of virus have been obtained by serial passage in infected cultures of mouse tumour cells. The first of these is the one described in the earlier parts of this paper, namely that isolated in Krebs 2 carcinoma cells, or K2 virus. The second strain of virus was isolated in mono-layer cultures of trypsinized cells from a solid mouse tumour (Sarcoma 180). Cells from this tumour are normally insusceptible to virus of mouse-brain origin, and the strain thus isolated appears to be a mutant. This mutant virus in its turn undergoes host-induced modification in the mouse brain after one growth cycle only. The phenotype of the modified virus is identical with that of mouse-brain virus; both are non-infective for Sarcoma 180 cells. Growth of the modified virus for one cycle only in Krebs 2 cells restores its infectivity for Sarcoma 180 cells, thus showing that this modification affects the virus phenotype while the genotype remains unaffected. Simi-larly one cycle of growth of mouse-brain virus in Krebs 2 cells does not endow this virus with infectivity for Sarcoma 180 cells. S180 and mouse-brain viruses thus possess different genotypes.

K2 virus in contrast does not undergo phenotypic modification in the mouse brain, and is therefore genotypically different from S180 virus. This is supported by the difference in plaque size, when these two viruses are plated under standard conditions. Moreover, a genetical difference between mouse brain and K2 virus is shown by their different effect upon Krebs 2 cells and also by their different plaque properties. These differences between the three strains of EMC virus are summarized in Table 1. Phenotypic modification of a virus by its host has previously been described by Luria (1953) and others with DNA viruses such as bacteriophages. The present observations show that host modification can also occur with an RNA virus.

In conclusion, these preliminary studies show that the use of ascites tumour cells in suspension enable us to study the intracellular behaviour of a small RNA virus, with the same precision as bacteriophage. Moreover, the same system is valuable for genetic studies. However, many questions remain as yet unanswered. For example, what is the intracellular site of

Table 1

	Mouse-brain virus	K2 virus	S180 virus
Cytopathic destruction of S180 cells	−	+	+
Cytopathic destruction of Krebs 2 cells	+	+	+
Phenotypic modification by mouse brain		−	+
Cytopathic destruction of S180 cells by mouse-brain phenotype after passage in Krebs 2 cells		−	+
Plaque size	Small	Large	Small

virus multiplication? So far it has been assumed that this is the cytoplasm, but RNA is found in the cell both in the nucleus and in the cytoplasm, and, assuming that there is some close liason between the virus RNA and that of its host, this could take place at either site. Thus we cannot reject *a priori* the participation of the cell nucleus in the synthesis of RNA viruses. In this connexion, Dunnebacke (1956) has described nucleolar degeneration as the first sign of cell degeneration induced by poliomyelitis in human amnion cells. The fact that the whole process of virus multiplication can be initiated by an amount of RNA equivalent to that contained in one microsome makes it likely that some key site in the cell is involved. Can this site be the nucleus? At the moment, however, speculation is unprofitable, since the EMC virus/ascites cell system provides the opportunity of solving many of these problems experimentally.

REFERENCES

ACKERMANN, W. W., RABSON, A. & KURTZ, H. (1954). *J. Exp. Med.* **100**, 437.

ADAMS, M. H. (1950). *Methods in Medical Research*, vol. 2, p. 6.

COLTER, J. S., BIRD, H. H. & BROWN, R. A. (1957). *Nature, Lond.* **179**, 859.

COOPER, P. D. (1955). *Virology*, **1**, 397.

DELBRUCK, M. (1949). In *Unités douées de continuité génétique.* C.N.R.S. Paris, pp. 33–4.

DULBECCO, R. (1955). *Physiol. Rev.* **35**, 301.

DULBECCO, R. & VOGT, M. (1954*a*). *J. Exp. Med.* **99**, 167.

DULBECCO, R. & VOGT, M. (1954*b*). *J. Exp. Med.* **99**, 183.

DUNNEBACKE, T. H. (1956). *Virology*, **2**, 811.

FRISCH-NIGGEMEYER, W. (1956). *Nature, Lond.* **178**, 307.

GARD, S. & HELLER, L. (1951). *Proc. Soc. Exp. Biol., N.Y.* **76**, 68.

GIERER, A. & SCHRAMM, G. (1956). *Nature, Lond.* **177**, 702.

HAMERS-CASTERMANN, C. & JEENER, C. A. (1957). *Virology*, **3**, 197.

HENLE, G., GIRARDI, A. & HENLE, W. (1955). *J. Exp. Med.* **101**, 25.

HERRIOTT, R. M. (1951). *J. Bact.* **61**, 752.

HERSHEY, A. D. (1957). *Advances in Virus Research* **4**, 25.

HERSHEY, A. D. & CHASE, M. (1952). *J. Gen. Physiol.* **36**, 39.

HINSHELWOOD, C. N. (1947). *The Chemical Kinetics of the Bacterial Cell.* Oxford University Press.

HINSHELWOOD, C. N. (1952). *J. Chem. Soc.* **136**, 795.

HORVATH, F. & JUNGEBLUT, C. (1952). *J. Immunol.* **68**, 627.

HOSKINS, J. M. & SANDERS, F. K. (1957). *Brit. J. Exp. Pathol.* **38**, 268.

HOSKINS, J. M., MEYNELL, G. G. & SANDERS, F. K. (1956). *Exp. Cell Res.* **11**, 297.

KLEIN, G. & KLEIN, E. (1951). *Cancer Res.* **11**, 466.

LANNI, F. & LANNI, Y. T. (1953). *Cold Spr. Harb. Symp. Quant. Biol.* **18**, 159.

LE CLERC, J. (1956). *Nature, Lond.* **177**, 578.

LURIA, S. I. (1953). *Cold Spr. Harb. Symp. Quant. Biol.* **18**, 237.

LWOFF, A., DULBECCO, R., VOGT, M. & LWOFF, M. (1955). *Virology*, **1**, 128.

PEREIRA, H. G. (1954). *J. Gen. Microbiol.* **10**, 500.

POLLOCK, M. A. (1953). *Symp. Soc. Gen. Microbiol.* **3**, 150.

RUBIN, H., BALUDA, M. & HOTCHIN, J. E. (1955). *J. Exp. Med.* **101**, 205.

RUBIN, H., FRANKLIN, R. M. & BALUDA, M. (1957). *Virology*, **3**, 587.

SANDERS, F. K. (1957). *Proc. R. Soc. Med.* **50**, 911.

SCHREK, R. (1936). *Amer. J. Cancer*, **28**, 389.

WATSON, J. D. & CRICK, F. H. C. (1953). *Cold Spr. Harb. Symp. Quant. Biol.* **18**, 123.

WECKER, VON E. & SCHÄFER, W. (1957). *Z. Naturf.* **12b**, 415.

WILLIAMS, R. C. (1957). In CIBA Foundation Symposium on *The Nature of Viruses*, p. 19. Ed. Wolstenholme, G. E. W. & Millar, E. C. P. London: Churchill.

ON PROTEIN SYNTHESIS

By F. H. C. CRICK

Medical Research Council Unit for the Study of Molecular Biology,
Cavendish Laboratory, Cambridge

I. INTRODUCTION

Protein synthesis is a large subject in a state of rapid development. To cover it completely in this article would be impossible. I have therefore deliberately limited myself here to presenting a broad general view of the problem, emphasizing in particular well-established facts which require explanation, and only selecting from recent work those experiments whose implications seem likely to be of lasting significance. Much very recent work, often of great interest, has been omitted because its implications are not clear. I have also tried to relate the problem to the other central problems of molecular biology—those of gene action and nucleic acid synthesis. In short, I have written for the biologist rather than the biochemist, the general reader rather than the specialist. More technical reviews have appeared recently by Borsook (1956), Spiegelman (1957), and Simkin & Work (1957b and this Symposium).

The importance of proteins

It is an essential feature of my argument that in biology proteins are uniquely important. They are not to be classed with polysaccharides, for example, which by comparison play a very minor role. Their nearest rivals are the nucleic acids. Watson said to me, a few years ago, 'The most significant thing about the nucleic acids is that we don't know what they do.' By contrast the most significant thing about proteins is that they can do almost anything. In animals proteins are used for structural purposes, but this is not their main role, and indeed in plants this job is usually done by polysaccharides. *The main function of proteins is to act as enzymes.* Almost all chemical reactions in living systems are catalysed by enzymes, and all known enzymes are proteins. It is at first sight paradoxical that it is probably easier for an organism to produce a new protein than to produce a new small molecule, since to produce a new small molecule one or more new proteins will be required in any case to catalyse the reactions.

I shall also argue that the main function of the genetic material is to control (not necessarily directly) the synthesis of proteins. There is a little direct evidence to support this, but to my mind the psychological drive behind this hypothesis is at the moment independent of such evidence.

Once the central and unique role of proteins is admitted there seems little point in genes doing anything else. Although proteins can *act* in so many different ways, the way in which they are *synthesized* is probably uniform and rather simple, and this fits in with the modern view that gene action, being based upon the nucleic acids, is also likely to be uniform and rather simple.

Biologists should not deceive themselves with the thought that some new class of biological molecules, of comparable importance to the proteins, remains to be discovered. This seems highly unlikely. In the protein molecule Nature has devised a unique instrument in which an underlying simplicity is used to express great subtlety and versatility; it is impossible to see molecular biology in proper perspective until this peculiar combination of virtues has been clearly grasped.

II. THE PROBLEM

Elementary facts about proteins

(1) *Composition.* Simple (unconjugated) proteins break down on hydrolysis to amino acids. There is good evidence that in a native protein the amino acids are condensed into long polypeptide chains. A typical protein, of molecular weight about 25,000, will contain some 230 residues joined end-to-end to form a single polypeptide chain.

Two points are important. First, the actual chemical step required to form the covalent bonds of the protein is always the same, irrespective of the amino acid concerned, namely the formation of the peptide link with the elimination of water. Apart from minor exceptions (such as S—S links and, sometimes, the attachment of a prosthetic group) *all* the covalent links within a protein are formed in this way. Covalently, therefore, a protein is to a large extent a linear molecule (in the topological sense) and there is little evidence that the backbone is ever branched. From this point of view the cross-linking by S—S bridges is looked upon as a secondary process.

The second important point—and I am surprised that it is not remarked more often—is that only about twenty different *kinds* of amino acids occur in proteins, and that these same twenty occur, broadly speaking, in *all* proteins, of whatever origin—animal, plant or micro-organism. Of course not every protein contains every amino acid—the amino acid tryptophan, which is one of the rarer ones, does not occur in insulin, for example— but the majority of proteins contain at least one of each of the twenty amino acids. In addition all these twenty amino acids (apart from glycine) have the L configuration when they occur in genuine proteins.

There are a few proteins which contain amino acids not found elsewhere—the hydroxyproline of collagen is a good example—but in all such cases it is possible to argue that their presence is due to a modification of the protein after it has been synthesized or to some other abnormality. In Table 1, I have listed the standard twenty amino acids believed to be of universal occurrence and also, in the last column, some of the exceptional ones. The assignment given in Table 1 might not be agreed by everyone,

Table 1

The magic twenty amino acids found universally in proteins

		Other amino acids found in proteins
Glycine	Asparagine	Hydroxyproline*
Alanine	Glutamine	Hydroxylysine
Valine	Aspartic acid	Phosphoserine
Leucine	Glutamic acid	Diaminopimelic acid
Isoleucine	Arginine	Thyroxine and related molecules
Proline*	Lysine	
Phenylalanine	Histidine	Cystine†
Tyrosine	Tryptophan	
Serine	Cysteine†	
Threonine	Methionine	

* These are, of course, imino acids. This distinction is not made in the text.
† This classification implies that all the cystine found in proteins is formed by the joining together of two cysteine molecules.

as the evidence is incomplete, but more agreement could be found for this version than for any other. Curiously enough this point is slurred over by almost all biochemical textbooks, the authors of which give the impression that they are trying to include as many amino acids in their lists as they can, without bothering to distinguish between the magic twenty and the others. (But see a recent detailed review by Synge, 1957.)

(2) *Homogeneity.* Not only is the composition of a given protein fixed, but we have every reason to believe that the exact order of the amino acid residues along the polypeptide chains is also rigidly determined: that each molecule of haemoglobin in your blood, for example, has exactly the same sequence of amino acids as every other one. This is clearly an overstatement; the mechanism must make mistakes sometimes, and, as we shall see, there are also interesting exceptions which are under genetic control. Moreover, it is quite easy, in extracting a protein, to modify some of the molecules slightly without affecting the others, so that the 'pure' protein may appear heterogeneous. The exact amount of 'microheterogeneity' of proteins is controversial (see the review by Steinberg & Mihalyi, 1957), but this should not blind one to the astonishing degree of homogeneity of most proteins.*

(3) *Structure.* In a native globular protein the polypeptide chain is not

* The γ-globulins and other antibody molecules are exceptions to these generalizations. They are probably heterogeneous in folding and possibly to some extent in composition.

fully extended but is thrown into folds and superfolds, maintained by weak physical bonds, and in some cases by covalent —S—S— links and possibly some others. This folding is also thought to be at least broadly the same for each copy of a particular protein, since many proteins can be crystallized, though the evidence for *perfect* homogeneity of folding is perhaps rather weak.* As is well known, if this folding is destroyed by heat or other methods the protein is said to be 'denatured'. The biological properties of most proteins, especially the catalytic action of enzymes, must depend on the exact spatial arrangement of certain side-groups on the surface of the protein, and altering this arrangement by unfolding the polypeptide chains will destroy the biological specificity of the proteins.

(4) *Amino acid requirements.* If one of the twenty amino acids is supplied to a cell it can be incorporated into proteins; amino acids are certainly protein precursors. The only exceptions are amino acids like hydroxyproline, which are not among the magic twenty. The utilization of peptides is controversial but the balance of evidence is against the occurrence of peptide intermediates. (See the discussion by Simkin & Work, this Symposium.)

If, for some reason, one of the twenty amino acids is not available to the organism, protein synthesis stops. Moreover, the continued synthesis of those parts of the protein molecules which do not contain that amino acid appears not to take place. This can be demonstrated particularly clearly in bacteria, but it is also true of higher animals. If a meal is provided that lacks an essential amino acid it is no use trying to make up for this deficiency by providing it a few hours later.

Very little is known about the accuracy with which the amino acids are selected. One would certainly expect, for example, that the mechanism would occasionally put a valine into an isoleucine site, but exactly how often this occurs is not known. The impression one gets from the rather meagre facts at present available is that mistakes occur rather infrequently.

In recent years it has been possible to introduce amino acid analogues into proteins by supplying the analogue under circumstances in which the amino acid itself is not easily available (see the review by Kamin & Handler, 1957). For example in *Escherichia coli* fluorophenylalanine has been incorporated in place of phenylalanine and tyrosine (Munier & Cohen, 1956) and it has even proved possible to replace completely the sulphur-containing amino acid methionine by its selenium analogue (Cohen & Cowie, 1957). Of the enzymes produced by the cell in these various ways some were active and some were inactive, as might have been expected.

(5) *Contrast with polysaccharides.* It is useful at this point to contrast proteins with polysaccharides to underline the differences between them.

* See previous footnote.

(I do not include nucleic acids among the polysaccharides.) Polysaccharides, too, are polymers, but each one is constructed from one, or at the most only about half-a-dozen kinds of monomer. Nevertheless many different monomers are found throughout Nature, some occurring here, some there. There is no standard set of monomers which is always used, as there is for proteins. Then polysaccharides are polydisperse—at least so far no monodisperse one has been found—and the order of their monomers is unlikely to be rigidly controlled, except in some very simple manner. Finally in those cases which have been carefully studied, such as starch, glycogen and hyaluronic acid, it has been found that the polymerization is carried out in a straightforward way by enzymes.

(6) *The genetics and taxonomy of proteins.* It is instructive to compare your own haemoglobin with that of a horse. Both molecules are indistinguishable in size. Both have similar amino acid compositions; similar but not identical. They differ a little electrophoretically, form different crystals, and have slightly different ends to their polypeptide chains. All these facts are compatible with their polypeptide chains having similar amino acid sequences, but with just a few changes here and there.

This 'family likeness' between the 'same' protein molecules *from different species* is the rule rather than the exception. It has been found in almost every case in which it has been looked for. One of the best-studied examples is that of insulin, by Sanger and his co-workers (Brown, Sanger & Kitai, 1955; Harris, Sanger & Naughton, 1956), who have worked out the complete amino acid sequences for five different species, only two of which (pig and whale) are the same. Interestingly enough the differences are all located in one small segment of one of the two chains.

Biologists should realize that before long we shall have a subject which might be called 'protein taxonomy'—the study of the amino acid sequences of the proteins of an organism and the comparison of them between species. It can be argued that these sequences are the most delicate expression possible of the phenotype of an organism and that vast amounts of evolutionary information may be hidden away within them.

There is, however, nothing in the evidence presented so far to prove that these differences between species are under the control of Mendelian genes. It could be argued that they were transmitted cytoplasmically through the egg. On the other hand, there is much evidence that genes do affect enzymes, especially from work on micro-organisms such as *Neurospora* (see Wagner & Mitchell, 1955). The famous 'one gene—one enzyme' hypothesis (Beadle, 1945) expresses this fact, although its truth is controversial (personally I believe it to be largely correct). However, in none of these cases has the protein (the enzyme, that is) ever been obtained pure.

There are a few cases where a Mendelian gene has been shown un-ambiguously to alter a protein, the most famous being that of human sickle-cell-anaemia haemoglobin, which differs electrophoretically from normal adult haemoglobin, as was discovered by Pauling and his co-workers (1949). Until recently it could have been argued that this was perhaps not due to a change in amino acid sequence, but only to a change in the folding. That the gene does in fact alter the amino acid sequence has now been con-clusively shown by my colleague, Dr Vernon Ingram. The difference is due to a valine residue occurring in the place of a glutamic acid one, and Ingram has suggestive evidence that this is the *only* change (Ingram, 1956, 1957). It may surprise the reader that the alteration of one amino acid out of a total of about 300 can produce a molecule which (when homozygous) is usually lethal before adult life but, for my part, Ingram's result is just what I expected.

The nature of protein synthesis

The basic dilemma of protein synthesis has been realized by many people, but it has been particularly aptly expressed by Dr A. L. Dounce (1956):

My interest in templates, and the conviction of their necessity, originated from a question asked me on my Ph.D. oral examination by Professor J. B. Sumner. He enquired how I thought proteins might be synthesized. I gave what seemed the obvious answer, namely, that enzymes must be responsible. Professor Sumner then asked me the chemical nature of enzymes, and when I answered that enzymes were proteins or contained proteins as essential components, he asked whether these enzyme proteins were synthesized by other enzymes and so on *ad infinitum*.

The dilemma remained in my mind, causing me to look for possible solutions that would be acceptable, at least from the standpoint of logic. The dilemma, of course, involves the specificity of the protein molecule, which doubtless depends to a considerable degree on the sequence of amino acids in the peptide chains of the protein. The problem is to find a reasonably simple mechanism that could account for specific sequences without demanding the presence of an ever-increasing number of new specific enzymes for the synthesis of each new protein molecule.

It is thus clear that the synthesis of proteins must be radically different from the synthesis of polysaccharides, lipids, co-enzymes and other small molecules; that it must be relatively simple, and to a considerable extent uniform throughout Nature; that it must be highly specific, making few mistakes; and that in all probability it must be controlled at not too many removes by the genetic material of the organism.

The essence of the problem

A systematic discussion of our present knowledge of protein synthesis could usefully be set out under three headings, each dealing with a flux:

the flow of energy, the flow of matter, and the flow of information. I shall not discuss the first of these here. I shall have something to say about the second, but I shall particularly emphasize the third—the flow of information.

By information I mean the specification of the amino acid sequence of the protein. It is conventional at the moment to consider separately the synthesis of the polypeptide chain and its folding. It is of course possible that there is a special mechanism for folding up the chain, but the more likely hypothesis is that the *folding is simply a function of the order of the amino acids*, provided it takes place as the newly formed chain comes off the template. I think myself that this latter idea may well be correct, though I would not be surprised if exceptions existed, especially the γ-globulins and the adaptive enzymes.

Our basic handicap at the moment is that we have no easy and precise technique with which to study how proteins are folded, whereas we can at least make some experimental approach to amino acid sequences. For this reason, if for no other, I shall ignore folding in what follows and concentrate on the determination of sequences. It is as well to realize, however, that the idea that the two processes can be considered separately is in itself an assumption.

The actual chemical step by which any two amino acids (or activated amino acids) are joined together is probably always the same, and may well not differ significantly from any other biological condensation. The unique feature of protein synthesis is that only a single standard set of twenty amino acids can be incorporated, and that for any particular protein *the amino acids must be joined up in the right order*. It is this problem, the problem of 'sequentialization', which is the crux of the matter, though it is obviously important to discover the exact chemical steps which lead up to and permit the crucial act of sequentialization.

As in even a small bacterial cell there are probably a thousand different kinds of protein, each containing some hundreds of amino acids in its own rigidly determined sequence, the amount of hereditary information required for sequentialization is quite considerable.

III. RECENT EXPERIMENTAL WORK

The role of the nucleic acids

It is widely believed (though not by every one) that the nucleic acids are in some way responsible for the control of protein synthesis, either directly or indirectly. The actual evidence for this is rather meagre. In the case of deoxyribonucleic acid (DNA) it rests partly on the T-even bacteriophages, since it has been shown, mainly by Hershey and his colleagues, that whereas the DNA of the infecting phage penetrates into the bacterial cell almost all the protein remains outside (see the review by Hershey, 1956); and also on Transforming Factor, which appears to be pure DNA, and which in at least one case, that of the enzyme mannitol phosphate dehydrogenase, controls the synthesis of a protein (Marmur & Hotchkiss, 1955). There is also the indirect evidence that DNA is the most constant part of the genetic material, and that genes control proteins. Finally there is the very recent evidence, mainly due to the work of Benzer on the rII locus of bacteriophage, that the functional gene—the 'cistron' of Benzer's terminology—consists of many sites arranged strictly *in a linear order* (Benzer, 1957) as one might expect if a gene controls the order of the amino acids in some particular protein.

As is well known, the correlation between ribonucleic acid (RNA) and protein synthesis was originally pointed out by Brachet and by Caspersson. Is there any more direct evidence for this connexion? In particular is there anything to support the idea that the sequentialization of the amino acids is controlled by the RNA?

The most telling evidence is the recent work on tobacco mosaic virus. A number of strains of the virus are known, and it is not difficult to show (since the protein sub-unit of the virus is small) that they differ in amino acid composition. Some strains, for example, have histidine in their protein, whereas others have none. Two very significant experiments have been carried out. In one, as first shown by Gierer & Schramm (1956), the RNA of the virus alone, although completely free of protein, appears to be infective, though the infectivity is low. In the other, first done by Fraenkel-Conrat, it has proved possible to separate the RNA from the protein of the virus and then recombine them to produce virus again. In this case the infectivity is comparatively high, though some of it is usually lost. If a recombined virus is made using the RNA of one strain and the protein of another, and then used to infect the plant, the new virus produced in the plant resembles very closely *the strain from which the RNA was taken*. If this strain had a protein which contained no histidine then the offspring will have no histidine either, although the plant had never been in contact with this particular protein before but only with the RNA from that strain. In other words *the viral RNA appears to carry at least part of the information which determines the composition of the viral protein*. Moreover the viral protein which was used to infect the cell was not copied to any appreciable extent (Fraenkel-Conrat, 1956).

It has so far not proved possible to carry out this experiment—a model of its kind—in any other system, although very recently it has been claimed that for two animal viruses the RNA alone appears to be infective.

Turnover experiments have shown that while the labelling of DNA is homogeneous that of RNA is not. The RNA of the cell is partly in the nucleus, partly in particles in the cytoplasm and partly as the 'soluble' RNA of the cell sap; many workers have shown that all these three fractions turn over differently. It is very important to realize in any discussion of the role of RNA in the cell that it is very inhomogeneous metabolically, and probably of more than one type.

The site of protein synthesis

There is no known case in Nature in which protein synthesis proper (as opposed to protein modification) occurs outside cells, though, as we shall see later, a certain amount of protein can probably be synthesized using broken cells and cell fragments. The first question to ask, therefore, is whether protein synthesis can take place in the nucleus, in the cytoplasm, or in both.

It is almost certain that protein synthesis can take place in the cytoplasm without the presence of the nucleus, and it is probable that it can take place to some extent in the nucleus by itself (see the review by Brachet & Chantrenne, 1956). Mirsky and his colleagues (see the review by Mirsky, Osawa & Allfrey, 1956) have produced evidence that some protein synthesis can occur in isolated nuclei, but the subject is technically difficult and in this review I shall quite arbitrarily restrict myself to protein synthesis in the cytoplasm.

In recent years our knowledge of the structure of the cytoplasm has enormously increased, due mainly to the technique of cutting thin sections for the electron microscope. The cytoplasm of many cells contains an 'endoplasmic reticulum' of double membranes, consisting mainly of protein and lipid (see the review of Palade, 1956). On one side of each membrane appear small electron-dense particles (Palade, 1955). Biochemical studies (Palade & Siekevitz, 1956; among others) have shown that these particles, which are about 100–200 Å. in diameter, consist almost entirely of protein and RNA, in about equal quantities. Moreover the major part of the RNA of the cell is found in these particles.

When such a cell is broken open and the contents fractionated by centrifugation, the particles, together with fragments of the endoplasmic reticulum, are found in the 'microsome' fraction, and for this reason I shall refer to them as microsomal particles.

These microsomal particles are found in almost all cells. They are particularly common in cells which are actively synthesizing protein, whereas the endoplasmic reticulum is most conspicuously present in (mammalian) cells which are secreting very actively. Thus both the cells of the pancreas and those of an ascites tumour contain large quantities of microsomal particles, but the tumour has little endoplasmic reticulum, whereas the pancreas has a lot. Moreover, there is no endoplasmic reticulum in bacteria.

On the other hand particles of this general description have been found in plant cells (Ts'o, Bonner & Vinograd, 1956), in yeast, and in various bacteria (Schachman, Pardee & Stanier, 1953); in fact in all cells which have been examined for them.

These particles have been isolated from various cells and examined in the ultra-centrifuge (Petermann, Mizen & Hamilton, 1952; Schachman *et al.* 1953; among others). The remarkable fact has emerged that they do not have a continuous distribution of sedimentation constants, but usually fall into several well-defined groups. Moreover some of the particles are probably simple aggregates of the others (Petermann & Hamilton, 1957). This uniformity suggests immediately that the particles, which have 'molecular weights' of a few million, have a definite structure. They are, in fact, reminiscent of the small spherical RNA-containing viruses, and Watson and I have suggested that they may have a similar type of substructure (Crick & Watson, 1956).

Biologists should contrast the older concept of *microsomes* with the more recent and significant one of *microsomal particles*. Microsomes came in all sizes, and were irregular in composition; microsomal particles occur in a few sizes only, have a more fixed composition and a much higher proportion of RNA. It was hard to identify microsomes in all cells, whereas RNA-rich particles appear to occur in almost every kind of cell. In short, microsomes were rather a mess, whereas microsomal particles appeal immediately to one's imagination. It will be surprising if they do not prove to be of fundamental importance.

It should be noted, however, that Simpson and his colleagues (Simpson & McLean, 1955; Simpson, McLean, Cohn & Brandt, 1957) have reported that protein synthesis can take place in mitochondria. It is known that mitochondria contain RNA, and it would be of great interest to know whether this RNA is in some kind of particle. Mitochondria are, of course, very widely distributed but they do not occur in lower forms such as bacteria. Similar remarks about RNA apply to the reported incorporation in chloroplasts (Stephenson, Thimann & Zamecnik, 1956).

Microsomal particles and protein synthesis

It has been shown by the use of radioactive amino acids that during protein synthesis the amino acids appear to flow through the microsomal particles. The most striking experiments are those of Zamecnik and his co-workers on the livers of growing rats (see the review by Zamecnik *et al.* 1956).

Two variations of the experiment were made. In the first the rat was given a rather large intravenous dose of a radioactive amino acid. After a predetermined time the animal was sacrificed, the liver extracted, its cells homogenized and the contents fractionated. It was found that the microsomal particle fraction was very rapidly labelled to a constant level.

In the second a very small shot of the radioactive amino acid was given, so that the liver received only a pulse of labelled amino acid, since this small amount was quickly used up. In this case the radioactivity of the microsomal particles rose very quickly *and then fell away*. Making plausible assumptions Zamecnik and his colleagues have shown that this behaviour is what one would expect if most of the protein of the microsomal particles was metabolically inert, but 1 or 2 % was turning over very rapidly, say within a minute or so.

Very similar results have been obtained by Rabinovitz & Olson (1956, 1957) using intact mammalian cells, in this case rabbit reticulocytes. They have also been able to show that the label passed into a well-defined globular protein, namely haemoglobin. Experiments along the same general lines have also been reported for liver by Simkin & Work (1957*a*).

We thus have direct experimental evidence that the microsomal particles are associated with protein synthesis, though the precise role they play is not clear.

Activating enzymes

It now seems very likely that the first step in protein synthesis is the activation of each amino acid by means of its special 'activating enzyme'. The activation requires ATP, and the evidence suggests that the reaction is

$$\text{amino acid} + \text{ATP} = \text{AMP} - \text{amino acid} + \text{pyrophosphate.}$$

The activated amino acid, which is probably a mixed anhydride of the form

$$R-\overset{\displaystyle NH_2}{\underset{\displaystyle H}{C}}-C\overset{\displaystyle O-\overset{\displaystyle O}{\overset{|}{P}}-O-\text{Ribose}-\text{Adenine}}{\underset{\displaystyle O}{\diagdown}} \overset{|}{O}$$

in which the carboxyl group of the amino acid is phosphorylated, appears to be tightly bound to its enzyme and is not found free in solution.

These enzymes were first discovered in the cell-sap fraction of rat liver cells by Hoagland (Hoagland, 1955; Hoagland, Keller & Zamecnik, 1956) and in yeast by Berg (1956). They have been shown by DeMoss & Novelli (1956) to be widely distributed in bacteria, and it is surmised that they occur in all cells engaged in protein synthesis. Recently Cole, Coote & Work (1957) have reported their presence in a variety of tissues from a number of animals.

So far good evidence has been found for this reaction for about half the standard twenty amino acids, but it is believed that further research will reveal the full set. Meanwhile Davie, Koningsberger & Lipmann (1956) have purified the tryptophan-activating enzyme. It is specific for tryptophan (and certain tryptophan analogues) and will only handle the L-isomer. Isolation of the tyrosine enzyme has also been briefly reported (Koningsberger, van de Ven & Overbeck, 1957; Schweet, 1957).

The properties of these enzymes are obviously of the greatest interest, and much work along these lines may be expected in the near future. For example, it has been shown that the tryptophan-activating enzyme contains what is probably a derivative of guanine (perhaps GMP) very tightly bound. It is possible to remove it, however, and to show that its presence is not necessary for the primary activation step. Since the enzyme is probably involved in the next step in protein synthesis it is naturally suspected that the guanine derivative is also required for this reaction, whatever it may be.

In vitro *incorporation*

In order to study the relationship between the activating enzymes and the microsomal particles it has proved necessary to break open the cells and work with certain partly purified fractions. Unfortunately it is rare to obtain substantial net protein synthesis from such systems, and there is a very real danger that the incorporation of the radioactivity does not represent true synthesis but is some kind of partial synthesis or exchange reaction. This distinction has been clearly brought out by Gale (1953). The work to be described, therefore, has to be accepted with reservations. (See the remarks of Simkin & Work, this Symposium.) It has been shown, however, in the work described below, that the amino acid is incorporated into true peptide linkage.

Again the significant results were first obtained by Zamecnik and his co-workers (reviewed in Zamecnik *et al.* 1956). The requirements so far known appear to fall into two parts:

(1) The activation of the amino acids for which, in addition to the labelled amino acid, one requires the 'pH 5' fraction, containing the activating enzymes, ATP and (usually) an ATP-generating system. There appears to be no requirement for any of the pyrimidine or guanine nucleotides.

(2) The transfer to the microsomal particles. For this one requires the previous system plus GTP or GDP (Keller & Zamecnik, 1956) and of

course the microsomal particles; the endoplasmic reticulum does not appear to be necessary (Littlefield & Keller, 1957).

Hultin & Beskow (1956) have reported an experiment which shows clearly that the amino acids become bound in some way. They first incubate the mixture described in (1) above. They then add a great excess of *un*labelled amino acid before adding the microsomal particles. Nevertheless some of the labelled amino acid is incorporated into protein, showing that it was in some place where it could not readily be diluted.

Very recently an intermediate reaction has been suggested by the work of Hoagland, Zamecnik & Stephenson (1957), who have discovered that in the first step the 'soluble' RNA contained in the 'pH 5' fraction became labelled with the radioactive amino acid. The bond between the amino acid and the RNA appears to be a covalent one. This labelled RNA can be extracted, purified, and then added to the microsomal fraction. In the presence of GTP the labelled amino acid is transferred from the soluble RNA to microsomal protein. This very exciting lead is being actively pursued.

Many other experiments have been carried out on cell-free systems, in particular by Gale & Folkes (1955) and by Spiegelman (see his review, 1957), but I shall not describe them here as their interpretation is difficult. It should be mentioned that Gale (reviewed in Gale, 1956) has isolated from hydrolysates of commercial-yeast RNA a series of fractions which greatly increase amino acid incorporation. One of them, the so-called 'glycine incorporation factor' has been purified considerably, and an attempt is being made to discover its structure.

RNA turnover and protein synthesis

From many points of view it seems highly likely that the *presence* of RNA is essential for cytoplasmic protein synthesis, or at least for specific protein synthesis. It is by no means clear, however, that the *turnover* of RNA is required.

In discussing this a strong distinction must be made between cells which are growing, and therefore producing new microsomal particles, and cells which are synthesizing without growth, and in which few new microsomal particles are being produced.

This is a difficult aspect of the subject as the evidence is to some extent conflicting. It appears reasonably certain that not *all* the RNA in the cytoplasm is turning over very rapidly—this has been shown, for example, by the Hokins (1954) working on amylase synthesis in slices of pigeon pancreas, though in the light of the recent work of Straub (this Symposium) the choice of amylase was unfortunate. On the other hand Pardee (1954) has demonstrated that mutants of *Escherichia coli* which require uracil or adenine cannot synthesize β-galactosidase unless the missing base is provided.

Can RNA be synthesized without protein being synthesized? This can be brought about by the use of chloramphenicol. In bacterial systems chloramphenicol stops protein synthesis dead, but allows 'RNA' synthesis to continue. A very interesting phenomenon has been uncovered in *E. coli* by Pardee & Prestidge (1956), and by Gros & Gros (1956). If a mutant is used which requires, say, leucine, then when the external supply of leucine is exhausted both protein and RNA synthesis cease. If now chloramphenicol is added there is no effect, but if in addition the cells are given a small amount of leucine then rapid RNA synthesis takes place. If the chloramphenicol is removed, so that protein synthesis restarts, then this leucine is built into proteins and then, once again, the synthesis of both protein and RNA is prevented. In other words it appears as if 'free' leucine (i.e. not bound into proteins) is required for RNA synthesis. This effect is not peculiar to leucine and has already been found for several amino acids and in several different organisms (Yčas & Brawerman, 1957).

As a number of people have pointed out, the most likely interpretation of these results is that protein and RNA require *common intermediates* for their synthesis, consisting in part of amino acids and in part of RNA components such as nucleotides. This is a most valuable idea; it explains a number of otherwise puzzling facts and there is some hope of getting close to it experimentally.

For completeness it should be stated that Anfinsen and his co-workers have some evidence that proteins are not produced from (activated) amino acids in a single step (see the review by Steinberg, Vaughan & Anfinsen, 1956), since they find unequal labelling between the same amino acid at different points on the polypeptide chain, but this interpretation of their results is not accepted by all workers in the field. This is discussed more fully by Simkin & Work (this Symposium).

Summary of experimental work

Both DNA and RNA have been shown to carry some of the specificity for protein synthesis. The RNA of almost all types of cell is found mainly in rather uniform, spherical, virus-like particles in the cytoplasm, known as microsomal particles. Most of their protein and RNA is metabolically rather inert. Amino acids, on their way into protein, have been shown to pass rapidly through these particles.

An enzyme has been isolated which, when supplied with tryptophan and ATP, appears to form an activated tryptophan. There is evidence that there exist similar enzymes for most of the other amino acids. These enzymes are widely distributed in Nature.

Work on cell fractions is difficult to interpret but suggests that the first step in protein synthesis involves these enzymes, and that the subsequent transfer of the activated amino acids to the microsomal particles requires GTP. The soluble RNA also appears to be involved in this process.

Whereas the presence of RNA is probably required for true protein synthesis its rapid turnover does not appear to be necessary, at least not for all the RNA. There is suggestive evidence that common intermediates, containing both amino acids and nucleotides, occur in protein synthesis.

IV. IDEAS ABOUT PROTEIN SYNTHESIS

It is an extremely difficult matter to present current ideas about protein synthesis in a stimulating form. Many of the general ideas on the subject have become rather stale, and an extended discussion of the more detailed theories is not suitable in a paper for non-specialists. I shall therefore restrict myself to an outline sketch of my own ideas on cytoplasmic protein synthesis, some of which have not been published before. Finally I shall deal briefly with the problem of 'coding'.

General principles

My own thinking (and that of many of my colleagues) is based on two general principles, which I shall call the Sequence Hypothesis and the Central Dogma. The direct evidence for both of them is negligible, but I have found them to be of great help in getting to grips with these very complex problems. I present them here in the hope that others can make similar use of them. Their speculative nature is emphasized by their names. It is an instructive exercise to attempt to build a useful theory without using them. One generally ends in the wilderness.

The Sequence Hypothesis

This has already been referred to a number of times. In its simplest form it assumes that the specificity of a piece of nucleic acid is expressed solely by the sequence of its bases, and that this sequence is a (simple) code for the amino acid sequence of a particular protein.

This hypothesis appears to be rather widely held. Its virtue is that it unites several remarkable pairs of generalizations: the central biochemical importance of proteins and the dominating biological role of genes, and in particular of their nucleic acid; the linearity of protein molecules (considered covalently) and the genetic linearity within the functional gene, as shown by the work of Benzer (1957) and Pontecorvo (this Symposium); the simplicity of the composition of protein molecules and the simplicity of the nucleic acids. Work is actively proceeding in several laboratories, including our own, in an attempt to provide more direct evidence for this hypothesis.

The Central Dogma

This states that once 'information' has passed into protein *it cannot get out again.* In more detail, the transfer of information from nucleic acid to nucleic acid, or from nucleic acid to protein may be possible, but transfer from protein to protein, or from protein to nucleic acid is impossible. Information means here the *precise* determination of sequence, either of bases in the nucleic acid or of amino acid residues in the protein.

This is by no means universally held—Sir Macfarlane Burnet, for example, does not subscribe to it—but many workers now think along these lines. As far as I know it has not been *explicitly* stated before.

Some ideas on cytoplasmic protein synthesis

From our assumptions it follows that there must be an RNA template in the cytoplasm. The obvious place to locate this is in the microsomal particles, because their uniformity of size suggests that they have a regular structure. It also follows that the synthesis of at least some of the microsomal RNA must be under the control of the DNA of the nucleus. This is because the amino acid sequence of the human haemoglobins, for example, is controlled at least in part by a Mendelian gene, and because spermatozoa contain no RNA. Therefore, granted our hypotheses, the information must be carried by DNA.

What can we guess about the structure of the microsomal particle? On our assumptions the protein component of the particles can have no significant role in determining the amino acid sequence of the proteins which the particles are producing. We therefore assume that their main function is a structural one, though the possibility of some enzyme activity is not excluded. The simplest model then becomes one in which each particle is made of the same protein, or proteins, as every other one in the cell, and has the same basic *arrangement* of the RNA, but that different particles have, in general, different base-sequences in their RNA, and therefore produce different proteins. This is exactly the type of structure found in tobacco mosaic virus, where the interaction between RNA and protein does not depend upon the sequence of bases of the RNA (Hart & Smith, 1956). In addition Watson and I have suggested (Crick & Watson, 1956), by analogy with the spherical viruses, that the protein of microsomal particles is probably made of many identical sub-units arranged with cubic symmetry.

On this oversimplified picture, therefore, the microsomal particles in a cell are all the same (except for the base-sequence of their RNA) and are metabolically rather inert. The RNA forms the template and the protein supports and protects the RNA.

This idea is in sharp contrast to what one would naturally assume at first glance, namely that the protein of the microsomal particles consists entirely of protein being synthesized. The surmise that most of the protein is structural was derived from considerations about the structure of virus particles and about coding; it was independent of the direct experimental evidence of Zamecnik and his colleagues that only a small fraction of the protein turns over rapidly, so that this agreement between theory and experiment is significant, as far as it goes.

It is obviously of the first importance to know how the RNA of the particles is arranged. It is a natural deduction from the Sequence Hypothesis that the RNA backbone will follow as far as possible a spatially regular path, in this case a helix, essentially because the fundamental operation of making the peptide link is always the same, and we therefore expect any template to be spatially regular.

Although we do not yet know the structure of isolated RNA (which may be an artifact) we do know that a pair of RNA-like molecules can under some circumstances form a double-helical structure, somewhat similar to DNA, because Rich & Davies (1956) have shown that when the two polyribotides, polyadenylic acid and polyuridylic acid (which have the same backbone as RNA) are mixed together they wind round one another to form a double helix, presumably with their bases paired. It would not be surprising, therefore, if the RNA backbone took up a helical configuration similar to that found for DNA.

This suggestion is in contrast to the idea that the RNA and protein interact in a complicated, irregular way to form a 'nucleoprotein'. As far as I know there is at the moment no direct experimental evidence to decide between these two points of view.

However, even if it turns out that the RNA is (mainly) helical and that the structural protein is made of sub-units arranged with cubic symmetry it is not at all obvious how the two could fit together. In abstract terms the problem is how to arrange a long fibrous object inside a regular polyhedron. It is for this reason that the structure of the spherical viruses is of great interest in this context, since we suspect that the same situation occurs there; moreover they are at the moment more amenable to experimental attack. A possible arrangement, for example, is one in which the axes of the RNA helices run radially and clustered in groups of five, though it is always possible that the arrangement of the RNA is irregular.

It would at least be of some help if the approximate location of the RNA in the microsomal particles could be discovered. Is it on the outside or the inside of the particles, for example, or even both? Is the microsomal particle a rather open structure, like a sponge, and if it is what size of molecule can diffuse in and out of it? Some of these points are now ripe for a direct experimental attack.

The adaptor hypothesis

Granted that the RNA of the microsomal particles, regularly arranged, is the template, how does it direct the amino acids into the correct order? One's first naïve idea is that the RNA will take up a configuration capable of forming twenty different 'cavities', one for the side-chain of each of the twenty amino acids. If this were so one might expect to be able to play the problem backwards—that is, to find the configuration of RNA by trying to form such cavities. All attempts to do this have failed, and on physical-chemical grounds the idea does not seem in the least plausible (Crick, 1957a). Apart from the phosphate-sugar backbone, which we have assumed to be regular and perhaps linked to the structural protein of the particles, RNA presents mainly a sequence of sites where hydrogen bonding could occur. One would expect, therefore, that whatever went on to the template in a *specific* way did so by forming hydrogen bonds. It is therefore a natural hypothesis that the amino acid is carried to the template by an 'adaptor' molecule, and that the adaptor is the part which actually fits on to the RNA. In its simplest form one would require twenty adaptors, one for each amino acid.

What sort of molecules such adaptors might be is anybody's guess. They might, for example, be proteins, as suggested by Dounce (1952) and by the Hokins (1954) though personally I think that proteins, being rather large molecules, would take up too much space. They might be quite unsuspected molecules, such as amino sugars. But there is one possibility which seems inherently more likely than any other—that they might contain nucleotides. This would enable them to join on to the RNA template by the same 'pairing' of bases as is found in DNA, or in polynucleotides.

If the adaptors were small molecules one would imagine that a separate enzyme would be required to join each adaptor to its own amino acid and that the specificity required to distinguish between, say, leucine, iso-leucine and valine would be provided by these enzyme molecules instead of by cavities in the RNA. Enzymes, being made of protein, can probably make such distinctions more easily than can nucleic acid.

An outline picture of the early stages of protein synthesis might be as follows: the template would consist of perhaps a single chain of RNA. (As far as we know a single isolated RNA backbone has no regular configuration (Crick, 1957*b*) and one has to assume that the backbone is supported in a helix of the usual type by the structural protein of the microsomal particles.) Alternatively the template might consist of a pair of chains. Each adaptor molecule containing, say, a di- or trinucleotide would each be joined to its own amino acid by a special enzyme. These molecules would then diffuse to the microsomal particles and attach to the proper place on the bases of the RNA by base-pairing, so that they would then be in a position for polymerization to take place.

It will be seen that we have arrived at the idea of common intermediates without using the direct experimental evidence in their favour; but there is one important qualification, namely that the nucleotide part of the intermediates must be specific for each amino acid, at least to some extent. It is not sufficient, from this point of view, merely to join adenylic acid to each of the twenty amino acids. Thus one is led to suppose that after the activating step, discovered by Hoagland and described earlier, some other more specific step is needed before the amino acid can reach the template.

The soluble RNA

If trinucleotides, say, do in fact play the role suggested here their synthesis presents a puzzle, since one would not wish to invoke too many enzymes to do the job. It seems to me plausible, therefore, that the twenty different adaptors may be synthesized by the *breakdown* of RNA, probably the 'soluble' RNA. Whether this is in fact the same action which the 'activating enzymes' carry out (presumably using GTP in the process) remains to be seen.

From this point of view the RNA with amino acids attached reported recently by Hoagland, Zamecnik & Stephenson (1957), would be a half-way step in this process of breaking the RNA down to trinucleotides and joining on the amino acids. Of course alternative interpretations are possible. For example, one might surmise that numerous amino acids become attached to this RNA and then proceed to polymerize, perhaps inside the microsomal particles. I do not like these ideas, because the supernatant RNA appears to be too short to code for a complete polypeptide chain, and yet too long to join on to template RNA (in the microsomal particles) by base-pairing, since it would take too great a time for a piece of RNA twenty-five nucleotides long, say, to diffuse to the correct place in the correct particles. If it were only a trinucleotide on the other hand, there would be many different 'correct' places for it to go to (wherever a valine was required, say), and there would be no undue delay.

Leaving theories on one side, it is obviously of the greatest interest to know what molecules actually pass from the 'pH 5 enzymes' to the microsomal particles. Are they small molecules, free in solution, or are they bound to protein? Can they be isolated? This seems at the moment to be one of the most fruitful points at which to attack the problem.

Subsequent steps

What happens after the common intermediates have entered the microsomal particles is quite obscure. Two views are possible, which might be called the Parallel Path and the Alternative Path theories. In the first an intermediate is used to produce both protein and RNA at about the same time. In the second it is used to produce either protein, or RNA, but not both. If we knew the exact nature of the intermediates we could probably decide which of the two was more likely. At the moment there seems little reason to prefer one theory to the other.

The details of the polymerization step are also quite unknown. One tentative theory, of the Parallel Path type, suggests that the intermediates first polymerize to give an RNA molecule with amino acids attached. This process removes it from the template and it diffuses outside the microsomal particle. There the RNA folds to a new configuration, and the amino acids become polymerized to form a polypeptide chain, which folds up as it is made to produce the finished protein. The RNA, now free of amino acids, is then broken down to produce fresh intermediates. A great variety of theories along these lines can be constructed. I shall not discuss these further here, nor shall I describe the various speculations about the actual details of the chemical steps involved.

Two types of RNA

It is an essential feature of these ideas that there should be *at least two types of RNA in the cytoplasm*. The first, which we may call 'template RNA' is located inside the microsomal particles. It is probably synthesized in the nucleus (Goldstein & Plaut, 1955) under the direction of DNA, and carries the information for sequentialization. It is metabolically inert during protein synthesis, though naturally it may show turnover whenever microsomal particles are being synthesized (as in growing cells), or breaking down (as in certain starved cells).

The other postulated type of RNA, which we may call 'metabolic RNA', is probably synthesized (from common intermediates) in the microsomal particles, where its sequence is determined by base-pairing with the template RNA. Once outside the microsomal particles it becomes 'soluble RNA' and is constantly being broken down to form the common intermediates with the amino acids. It is also possible that some of the soluble RNA may be synthesized in a random manner in the cytoplasm; perhaps, in bacteria, by the enzyme system of Grunberg-Manago & Ochoa (1955).

One might expect that there would also be metabolic RNA in the nucleus. The existence of these different kinds of RNA may well explain the rather conflicting data on RNA turnover.

The coding problem

So much for biochemical ideas. Can anything about protein synthesis be discovered by more abstract arguments? If, as we have assumed, the sequence of bases along the nucleic acid determines the sequence of amino acids of the protein being synthesized, it is not unreasonable to suppose that this inter-relationship is a simple one, and to invent abstract descriptions of it. This problem of how, in outline, the sequence of four bases 'codes' the sequence of the twenty amino acids is known as the coding problem. It is regarded as being independent of the biochemical steps involved, and deals only with the transfer of information.

This aspect of protein synthesis appeals mainly to those with a background in the more sophisticated sciences. Most biochemists, in spite of being rather fascinated by the problem, dislike arguments of this kind. It seems to them unfair to construct theories without adequate experimental facts. Cosmologists, on the other hand, appear to lack such inhibitions.

The first scheme of this kind was put forward by Gamow (1954). It was supposedly based on some features of the structure of DNA, but these are irrelevant. The essential features of Gamow's scheme were as follows:

(a) Three bases coded one amino acid.

(b) Adjacent triplets of bases overlapped. See Fig. 1.

(c) More than one triplet of bases stood for a particular amino acid (degeneracy).

In other words it was an overlapping degenerate triplet code. Such a code imposes severe restrictions on the amino acid sequences it can produce. It is quite easy to disprove Gamow's code from a study of known sequences—even the sequences of the insulin molecule are sufficient. However, there are a very large number of codes of this general type. It might be thought almost impossible to disprove them all without enumerating them, but this has recently been done by Brenner (1957), using a neat argument. He has shown that the reliable amino acid sequences already known are enough to make *all* codes of this type impossible.

Attempts have been made to discover whether there are any obvious restrictions on the allowed amino acid sequences, although the sequence data available are very meagre (see the review by Gamow, Rich & Yčas,

1955). So far none has been found, and the present feeling is that it may well be that none exists, and that any sequence whatsoever can be produced. This is very far from being established, however, and for all we know there may be quite severe restrictions on the neighbours of the rarer amino acids, such as tryptophan.

If there is indeed a relatively simple code, then one of the most important biological constants is what Watson and I have called 'the coding ratio' (Crick & Watson, 1956). If B consecutive bases are required to code A consecutive amino acids, the coding ratio is the number B/A, when B and A are large. Thus in Gamow's code its value is unity, since a string of 1000 bases, for example, could code 998 amino acids. (Notice that when the coding ratio is greater than unity stereochemical problems arise, since a polypeptide chain has a distance of only about $3\frac{1}{2}$ Å. between its residues, which is about the minimum distance between successive bases in nucleic acid. However, it has been pointed out by Brenner (personal communication), that this difficulty may not be serious if the polypeptide chain leaves the template as it is being synthesized.)

```
                          B   C   A   C   D   D   A   B   A   B   D   C
                         ⎧ B   C   A
Overlapping code         ⎨     C   A   C
                         ⎩         A   C   D
                                       C   D   D

                         ⎧ B   C   A
Partial overlapping code ⎨     A   C   D
                         ⎩             D   D   A
                                           A   B   A

                         ⎧ B   C   A
Non-overlapping code     ⎨         C   D   D
                         ⎩                 A   B   A
                                               B   D   C
```

Fig. 1. The letters A, B, C, and D stand for the four bases of the four common nucleotides. The top row of letters represents an imaginary sequence of them. In the codes illustrated here each set of three letters represents an amino acid. The diagram shows how the first four amino acids of a sequence are coded in the three classes of codes.

If the code were of the non-overlapping type (see Fig. 1) one would still require a triplet of bases to code for each amino acid, since pairs of bases would only allow $4 \times 4 = 16$ permutations, though a possible but not very likely way round this has been suggested by Dounce, Morrison & Monty (1955). The use of triplets raises two difficulties. First, why are there not $4 \times 4 \times 4 = 64$ different amino acids? Second, how does one know which of the triplets to read (assuming that one doesn't start at an end)? For example, if the sequence of bases is ..., ABA, CDB, BCA, ACC, ..., where A, B, C and D represent the four bases, and where ABA is supposed to code one amino acid, CDB another one, and so on, how could one read it correctly if the commas were removed?

Very recently Griffith, Orgel and I have suggested an answer to both these difficulties which is of some interest because it *predicts* that there should be only twenty kinds of amino acid in protein (Crick, Griffith & Orgel, 1957). Gamow & Yčas (1955) had previously put forward a code with this property, known as the 'combination code' but the physical assumptions underlying their code lack plausibility. We assumed that some of the triplets (like *ABA* in the example above) correspond to an amino acid—make 'sense' as we would say—and some (such as *BAC* and *ACD*, etc., above) do not so correspond, or as we would say, make 'nonsense'.

We asked ourselves how many amino acids we could code if we allowed all possible sequences of amino acids, and yet never accidentally got 'sense' when reading the wrong triplets, that is those which included the imaginary commas. We proved that the upper limit is twenty, and moreover we could write down several codes which did in fact code twenty things. One such code of twenty triplets, written compactly is

$$A \quad B \quad A \atop B \qquad {A \atop B} \, C \, {A \atop {B \atop C}} \qquad {A \atop {B \atop C}} \, D \, {A \atop {B \atop {C \atop D}}}$$

where $A B {A \atop B}$ means that two of the allowed triplets are *ABA* and *ABB*, etc. The example given a little further back has been constructed using this code. You will see that *ABA, CDB, BCA* and *ACC* are among the allowed triplets, whereas the false overlapping ones in that example, such as *BAC, ACD* and *DBB*, etc., are not. The reader can easily satisfy himself that no sequence of these allowed triplets will ever give one of the allowed triplets in a false position. There are many possible mechanisms of protein synthesis for which this would be an advantage. One of them is described in our paper (Crick *et al.* 1957).

Thus we have deduced the magic number, twenty, in an entirely natural way from the magic number four. Nevertheless, I must confess that I find it impossible to form any considered judgment of this idea. It may be complete nonsense, or it may be the heart of the matter. Only time will show.

V. CONCLUSIONS

I hope I have been able to persuade you that protein synthesis is a central problem for the whole of biology, and that it is in all probability closely related to gene action. What are one's overall impressions of the present state of the subject? Two things strike me particularly. First, the existence of general ideas covering wide aspects of the problem. It is remarkable that one can formulate principles such as the Sequence Hypothesis and the

Central Dogma, which explain many striking facts and yet for which proof is completely lacking. This gap between theory and experiment is a great stimulus to the imagination. Second, the extremely active state of the subject experimentally both on the genetical side and the biochemical side. At the moment new and significant results are being reported every few months, and there seems to be no sign of work coming to a standstill because experimental techniques are inadequate. For both these reasons I shall be surprised if the main features of protein synthesis are not discovered within the next ten years.

It is a pleasure to thank Dr Sydney Brenner, not only for many interesting discussions, but also for much help in redrafting this paper.

REFERENCES

BEADLE, G. M. (1945). *Chem. Rev.* **37**, 15.
BENZER, S. (1957). In *The Chemical Basis of Heredity*. Ed. McElroy, W. D. & Glass, B. Baltimore: Johns Hopkins Press.
BERG, P. (1956). *J. Biol. Chem.* **222**, 1025.
BORSOOK, H. (1956). *Proceedings of the Third International Congress of Biochemistry, Brussels, 1955* (C. Liébecq, editor). New York: Academic Press.
BRACHET, J. & CHANTRENNE, H. (1956). *Cold Spr. Harb. Symp. Quant. Biol.* **21**, 329.
BRENNER, S. (1957). *Proc. Nat. Acad. Sci., Wash.* **43**, 687.
BROWN, N. H., SANGER, F. & KITAI, R. (1955). *Biochem. J.* **60**, 556.
COHEN, G. N. & COWIE, D. B. (1957). *C.R. Acad. Sci., Paris*, **244**, 680.
COLE, R. D., COOTE, J. & WORK, T. S. (1957). *Nature, Lond.* **179**, 199.
CRICK, F. H. C. (1957*a*). In *The Structure of Nucleic Acids and Their Role in Protein Synthesis*, p. 25. Cambridge University Press.
CRICK, F. H. C. (1957*b*). In *Cellular Biology, Nucleic Acids and Viruses*, 1957. New York Academy of Sciences.
CRICK, F. H. C., GRIFFITH, J. S. & ORGEL, L. E. (1957). *Proc. Nat. Acad. Sci., Wash.* **43**, 416.
CRICK, F. H. C. & WATSON, J. D. (1956). Ciba Foundation Symposium on *The Nature of Viruses*. London: Churchill.
DAVIE, E. W., KONINGSBERGER, V. V. & LIPMANN, F. (1956). *Arch. Biochem. Biophys.* **65**, 21.
DEMOSS, J. A. & NOVELLI, G. D. (1956). *Biochim. Biophys. Acta*, **22**, 49.
DOUNCE, A. L. (1952). *Enzymologia*, **15**, 251.
DOUNCE, A. L. (1956). *J. Cell. Comp. Physiol.* **47**, suppl. 1, 103.
DOUNCE, A. L., MORRISON, M. & MONTY, K. J. (1955). *Nature, Lond.* **176**, 597.
FRAENKEL-CONRAT, H. (1956). *J. Amer. Chem. Soc.* **78**, 882.
GALE, E. F. (1953). *Advanc. Protein Chem.* **8**, 283.
GALE, E. F. (1956). *Proceedings of the Third International Congress of Biochemistry, Brussels, 1955* (C. Liébecq, editor). New York: Academic Press.
GALE, E. F. & FOLKES, J. (1955). *Biochem. J.* **59**, 661, 675 and 730.
GAMOW, G. (1954). *Nature, Lond.* **173**, 318.
GAMOW, G., RICH, A. & YČAS, M. (1955). *Advanc. Biol. Med. Phys.* **4**. New York: Academic Press.

GAMOW, G. & YČAS, M. (1955). *Proc. Nat. Acad. Sci., Wash.* **41**, 1101.

GIERER, A. & SCHRAMM, G. (1956). *Z. Naturf.* **11**b, 138; also *Nature, Lond.* **177**, 702.

GOLDSTEIN, L. & PLAUT, W. (1955). *Proc. Nat. Acad. Sci., Wash.* **41**, 874.

GROS, F. & GROS, F. (1956). *Biochim. Biophys. Acta,* **22**, 200.

GRUNBERG-MANAGO, M. & OCHOA, S. (1955). *J. Amer. Chem. Soc.* **77**, 3165.

HARRIS, J. I., SANGER, F. & NAUGHTON, M. A. (1956). *Arch. Biochem. Biophys.* **65**, 427.

HART, R. G. & SMITH, J. D. (1956). *Nature, Lond.* **178**, 739.

HERSHEY, A. D. (1956). *Advances in Virus Research.* IV. New York: Academic Press.

HOAGLAND, M. B. (1955). *Biochim. Biophys. Acta,* **16**, 288.

HOAGLAND, M. B., KELLER, E. B. & ZAMECNIK, P. C. (1956). *J. Biol. Chem.* **218**, 345.

HOAGLAND, M. B., ZAMECNIK, P. C. & STEPHENSON, M. L. (1957). *Biochim. Biophys. Acta,* **24**, 215.

HOKIN, L. E. & HOKIN, M. R. (1954). *Biochim. Biophys. Acta,* **13**, 401.

HULTIN, T. & BESKOW, G. (1956). *Exp. Cell Res.* **11**, 664.

INGRAM, V. M. (1956). *Nature, Lond.* **178**, 792.

INGRAM, V. M. (1957). *Nature, Lond.* **180**, 326.

KAMIN, H. & HANDLER, P. (1957). *Annu. Rev. Biochem.* **26**, 419.

KELLER, E. B. & ZAMECNIK, P. C. (1956). *J. Biol. Chem.* **221**, 45.

KONINGSBERGER, V. V., VAN DE VEN, A. M. & OVERBECK, J. TH. G. (1957). *Proc. K. Akad. Wet. Amst.* B, **60**, 141.

LITTLEFIELD, J. W. & KELLER, E. B. (1957). *J. Biol. Chem.* **224**, 13.

MARMUR, J. & HOTCHKISS, R. D. (1955). *J. Biol. Chem.* **214**, 383.

MIRSKY, A. E., OSAWA, S. & ALLFREY, V. G. (1956). *Cold Spr. Harb. Symp. Quant. Biol.* **21**, 49.

MUNIER, R. & COHEN, G. N. (1956). *Biochim. Biophys. Acta,* **21**, 592.

PALADE, G. E. (1955). *J. Biochem. Biophys. Cytol.* **1**, 1.

PALADE, G. E. (1956). *J. Biochem. Biophys. Cytol.* **2**, 85.

PALADE, G. E. & SIEKEVITZ, P. (1956). *J. Biochem. Biophys. Cytol.* **2**, 171.

PARDEE, A. B. (1954). *Proc. Nat. Acad. Sci., Wash.* **40**, 263.

PARDEE, A. B. & PRESTIDGE, L. S. (1956). *J. Bact.* **71**, 677.

PAULING, L., ITANO, H. A., SINGER, S. J. & WELLS, I. C. (1949). *Science,* **110**, 543.

PETERMANN, M. L. & HAMILTON, M. G. (1957). *J. Biol. Chem.* **224**, 725; also *Fed. Proc.* **16**, 232.

PETERMANN, M. L., MIZEN, N. A. & HAMILTON, M. G. (1952). *Cancer Res.* **12**, 373.

RABINOVITZ, M. & OLSON, M. E. (1956). *Exp. Cell Res.* **10**, 747.

RABINOVITZ, M. & OLSON, M. E. (1957). *Fed. Proc.* **16**, 235.

RICH, A. &. DAVIES, D. R. (1956). *J. Amer. Chem. Soc.* **78**, 3548.

SCHACHMAN, H. K., PARDEE, A. B. & STANIER, R. Y. (1953). *Arch. Biochem. Biophys.* **43**, 381.

SCHWEET, R. (1957). *Fed. Proc.* **16**, 244.

SIMKIN, J. L. & WORK, T. S. (1957a). *Biochem. J.* **65**, 307.

SIMKIN, J. L. & WORK, T. S. (1957b). *Nature, Lond.* **179**, 1214.

SIMPSON, M. V. & McLEAN, J. R. (1955). *Biochim. Biophys. Acta,* **18**, 573.

SIMPSON, M. V., McLEAN, J. R., COHN, G. I. & BRANDT, I. K. (1957). *Fed. Proc.* **16**, 249.

SPIEGELMAN, S. (1957). In *The Chemical Basis of Heredity.* Ed. McElroy, W. D. & Glass, B. Baltimore: Johns Hopkins Press.

STEINBERG, D. & MIHALYI, E. (1957). In *Annu. Rev. Biochem.* **26**, 373.

STEINBERG, D., VAUGHAN, M. & ANFINSEN, C. B. (1956). *Science,* **124**, 389.

Stephenson, M. L., Thimann, K. V. & Zamecnik, P. C. (1956). *Arch. Biochem. Biophys.* **65**, 194.

Synge, R. (1957). In *The Origin of Life on the Earth.* U.S.S.R. Acad. of Sciences.

Ts'o, P. O. B., Bonner, J. & Vinograd, J. (1956). *J. Biochem. Biophys. Cytol.* **2**, 451.

Wagner, R. P. & Mitchell, H. K. (1955). *Genetics and Metabolism.* New York: John Wiley and Sons.

Yčas, M. & Brawerman, G. (1957). *Arch. Biochem. Biophys.* **68**, 118.

Zamecnik, P. C., Keller, E. B., Littlefield, J. W., Hoagland, M. B. & Loftfield, R. B. (1956). *J. Cell. Comp. Physiol.* **47**, suppl. 1, 81.

PROTEIN SYNTHESIS AS PART OF THE PROBLEM OF BIOLOGICAL REPLICATION

By J. L. SIMKIN and T. S. WORK

National Institute for Medical Research, Mill Hill, London, N.W. 7

An organism such as *Rhodospirillum rubrum* will grow in an inorganic salt medium. During growth the medium becomes progressively richer in proteins, in nucleic acids, in fats, in co-enzymes and in other intermediary metabolites. It would appear, superficially at least, that each type of molecule is undergoing reproduction, yet, in the absence of intact cells, there is no reproduction. In other words, it is the whole cell which is the self-reproducing unit and neither viruses nor genes, neither nucleic acids nor proteins can reproduce in the absence of the whole cell. Nevertheless, sufficient is now known of the functions of these various organic molecules to enable us to suggest that life might exist without polysaccharides or perhaps even without fats, but that no cell would be likely to exist in the absence of proteins since these substances provide the essential catalytically active surfaces without which metabolism cannot proceed.

It is tempting to assume that when the mechanism of protein biosynthesis is understood the mechanism of cell reproduction will be intelligible. An intact cell is, however, a highly organized structure, and it maintains within its envelope a whole series of balanced reactions, so that disturbance of any one of these may destroy the capacity for replication. In other words, replication requires the association and organization of many different types of molecule. It has been suggested at times that the gene is the key to biological replication but as pointed out by Lindegren (1955) the gene, which seems to serve the function of controller in the modern cell, may itself be a product of evolution.

This somewhat pessimistic note is a necessary reminder of the difficulties of the problem, but encouragement may be derived from other considerations. The cell as we know it could hardly have been suddenly created and, at some time during geological history, catalytically active molecules may have been reproduced without the need for cell structure. There is, of course, no assurance that such primitive methods of replication, which presumably must have started as a chance occurrence, are retained in the modern cell where replication involves the integration of so many different reactions. It is well to remember, however, that despite morphological evolution cells have retained a large measure of similarity in their biochemical reactions.

It is thus perhaps not too much to hope that the chemical principles involved in the replication of macromolecules by the modern cell are still fundamentally the same as those which were used by its earliest ancestors and that they are relatively simple. Proteins are undoubtedly the universal catalysts of the modern cell and knowledge of the biosynthetic mechanisms involved in protein synthesis, whatever their degree of complexity, may enable us to understand the chemical principles involved in replication, even though the process in its modern cellular form may be infinitely more complex.

There has existed for many years a school of thought which believes that proteins have no exact structure. The evidence for this view has been reviewed by Colvin, Smith & Cook (1954). There seems little doubt, in the face of current evidence, that this view will have to be abandoned and that proteins will have to be regarded in future as compounds with just as definite structures as sterols or porphyrins. The arguments in favour of this view have recently been ably summarized by Steinberg & Mihalyi (1957). Since proteins show such characteristics as structural specificity and catalytic activity, it is not sufficient to study the formation of undefined proteinaceous material. This requirement introduces one of the main experimental difficulties encountered in this field of study: until now it has not been possible to obtain, with certainty, the synthesis of specific and biologically active proteins except when using intact-cell preparations. Thus, for example, Borsook, Deasy, Haagen-Smit, Keighley & Lowy (1952) have failed to obtain haemoglobin synthesis in lysed reticulocytes although synthesis proceeds readily in the intact cell. Askonas has shown also that lymph gland slices from an immunized animal can synthesize antibody but that even slight damage of cell structure results in complete stoppage of antibody synthesis (Askonas, Simkin & Work, 1957). On the other hand, Gale & Folkes (1955) have reported increases in the activity of several enzymes in preparations of disrupted staphylococci. Straub and his colleagues (this Symposium) have studied amylase synthesis in pancreas homogenates and seemed at first to obtain synthesis of enzyme in cell-free preparations, but there is now some doubt as to whether this is so. Parke (unpublished), working at Mill Hill, has carried out a large series of experiments on pancreas homogenates similar to those reported by Straub, but although amylase synthesis in tissue slices was readily obtained no definite evidence could be obtained of synthesis in homogenates. There is thus an element of doubt as to what significance can be attached to results obtained when measuring the incorporation of amino acids into the mixed proteins of a tissue homogenate. Nevertheless, if the possibility of artifacts be kept in mind, the study of cell-free preparations may lead ultimately to an understanding of the mechanism of protein biosynthesis.

I. INTERMEDIATE STAGES IN PROTEIN SYNTHESIS

(a) The activation of amino acids

It has been clear for some years now that the formation of the peptide bond is an endergonic process and experiments on the mechanism of synthesis of simple amide bonds such as those of glutamine, hippuric acid, glutathione and pantothenic acid have strongly indicated that peptide bond formation is likely to involve carboxyl activation of amino acids. As already outlined by Crick (this Symposium), two enzymes have now been purified which catalyse carboxyl activation of tryptophan and tyrosine, and in addition there is sufficient evidence to suggest that there may be a separate activating enzyme for each amino acid. The product of reaction between activating enzyme, amino acid and adenosine triphosphate is an amino acid-adenosine monophosphate anhydride. There is as yet no direct proof that such anhydrides are intermediates in protein biosynthesis, but circumstantial evidence is suggestive. Thus, Sharon & Lipmann (1957), report that the tryptophan-activating enzyme is capable also of activating structural analogues of tryptophan such as tryptazan and azatryptophan but that it does not activate 5-methyl or 6-methyl tryptophan. The latter compounds inhibit the activating enzyme and immediately prevent bacterial growth, whereas the compounds activated are capable of substituting for tryptophan in protein synthesis in bacteria and permit limited growth.

(b) Small peptides as intermediates

Since proteins generally consist of long peptide chains containing 100 or more amino acids linked together in peptide linkage, it was a very reasonable first assumption that the long chains would be synthesized from amino acids by preliminary formation of small peptides followed by linkage together of suitable peptide sequences. In pursuance of this idea various laboratories have tried to isolate peptides from cell extracts. Only one peptide, glutathione, has been found in quantity but this tripeptide differs in structure from any peptide so far obtained by protein hydrolysis in that the linkage between glutamic acid and cysteinylglycine is a γ-linkage. Small quantities of peptides have been found in *Torula utilis* (Turba & Esser, 1955) and in *Pseudomonas hydrophila* (Connell & Watson, 1957) but there is considerable uncertainty as to whether these peptides are really intermediates in protein synthesis. In the interpretation of any results relating to the isolation of peptides from cell extracts, it is not sufficient to show that peptides are present or even to show that peptides are being rapidly synthesized from amino acids, as has been shown by Turba & Esser:

it must also be shown that these peptides are incorporated into proteins without first undergoing degradation to amino acids. Many attempts have been made to demonstrate direct utilization of peptides for bacterial growth, but the general trend of evidence is strongly in favour of the view that when peptides are used by bacteria as a source of nitrogen they are first degraded to amino acids (e.g. Meinhart & Simmonds, 1955).

An attempt has been made also to demonstrate direct utilization of peptide for casein synthesis. A goat was given intravenously a mixture of peptides produced by partial hydrolysis of goat casein but analysis of the milk secreted immediately following administration of the peptide mixture showed that no peptide had been directly used for casein synthesis (Godin & Work, 1956).

Reasonable objections can be raised to the evidence that cells do not use peptides. It may be that peptides present in the medium bathing the outside of a cell are unable to reach the site of protein synthesis. If, however, protein synthesis did occur in a stepwise fashion through peptides, then protein formed immediately after exposure of cells to radioactive amino acids would be expected to show uneven distribution of radioactive label along the peptide chain of newly formed protein. Evidence on this point is somewhat conflicting. Askonas, Campbell, Godin & Work (1955), measured the distribution of radioactivity in β-lactoglobulin and in casein from milk formed after administration of radioactive amino acids and found a uniform distribution of radioactivity between different portions of the protein molecule. Velick and his colleagues followed the synthesis of adolase, phosphorylase and glyceraldehyde-3-phosphate dehydrogenase in muscle, and their results were consistent with these findings (Heimberg & Velick, 1954; Simpson, 1955; Simpson & Velick, 1954). On the other hand, Steinberg & Anfinsen (1952), and Vaughan & Anfinsen (1954), studied the synthesis of ovalbumin, ribonuclease and insulin in tissue minces, under conditions where net synthesis of protein was not demonstrated. They found non-uniform distribution of isotope between different parts of the same molecule, a result which could be readily explained by the existence of peptide intermediates. This apparent contradiction between results obtained using intact animals and tissue minces is at present unresolved. Steinberg, Vaughan & Anfinsen (1956), after careful consideration of the results, conclude that none of the evidence is decisive and that their own results, which gave evidence of uneven labelling of proteins, could be explained without assuming the existence of peptides as intermediates in protein synthesis.

(c) The incorporation of amino acids into a trichloroacetic acid-insoluble fraction

It is easy to show incorporation of amino acids into newly formed protein in the intact cell. Various estimates have been made of the time required for re-appearance of amino acid in complete protein molecules. These estimates vary from a minute or less up to twenty minutes or more. The weight of evidence is now in favour of a very short time interval but the question remains: what are the individual stages in the process? The failure to find evidence for small peptides as intermediaries in protein synthesis has made it imperative to trace the fate of ingested amino acid during the short period before it re-appears as part of a complete protein molecule. Both Hultin (1950), and Borsook *et al.* (1950), administered isotopically labelled amino acids to normal animals, killed the animals shortly after-wards and, after homogenizing the liver, they fractionated the resultant material by differential centrifugation. In each case a preponderance of isotope was found in the protein of the microsome fraction. Shortly thereafter, Siekevitz (1952) showed that fairly extensive incorporation of radioactive amino acids into protein took place also in liver homogenates and that the radioactivity was found predominantly in the microsome fraction.

Incorporation can be obtained not only with whole homogenates but also with a preparation of microsomes suspended in cell sap or even, as pointed out by Crick (this Symposium), with isolated microsomal particles sus-pended in a partially purified mixture of activating enzymes (provided that the system is supplemented with adenosine triphosphate and guanosine triphosphate). It is now fully established, largely as a result of the work of Zamecnik, Keller, Littlefield, Hoagland & Loftfield (1956), that the first step in amino acid incorporation into microsomal protein in such systems is the activation of amino acids by specific activating enzymes plus adeno-sine triphosphate. There is evidence that the activated amino acids are first attached to a soluble ribonucleic acid (SRNA) before being transferred to the microsomal particles (Hoagland, Zamecnik & Stephenson, 1957). This SRNA is non-dialysable and the bond between the amino acid and the RNA is much more stable than that of the activated amino acid (anhydride) and indeed its stability is consistent with the assumption of an ester rather than an anhydride linkage. Transfer of amino acid from SRNA to micro-somal particle requires the presence of guanosine triphosphate and the final product, the proteinaceous material of the microsome particle, has been shown to contain isotopically labelled amino acids in true α-peptide linkage (Zamecnik *et al.* 1956). Thus the evidence obtained using homogenates suggests that protein synthesis proceeds by a series of discrete steps, some of which are more robust than the system for the synthesis of complete protein molecules, and each of which is susceptible to study in cell-free systems. Since, however, as already emphasized, synthesis of complete protein has not been demonstrated in homogenates and since all incorporation experiments carried out using isolated microsome systems involve isolation of trichloroacetic acid-insoluble proteinaceous material, it has become a matter of considerable importance to determine the nature of the labelled material produced in a cell-free system and to decide whether this is indeed a part of the protein-synthesizing system.

(d) *Comparison between microsome material labelled* in vivo *and* in vitro

As already mentioned, microsome material isolated from livers of animals killed shortly after an injection of radioactive amino acid is more active than any other cell fraction. Allfrey, Daly & Mirsky (1953) showed that the pancreatic microsomal proteins were not uniformly labelled and that treatment with ribonuclease caused the release of a protein with higher isotope content than the bulk of the material. Simkin (1955) found that liver microsomal preparations could be fractionated by successive extractions with solutions of differing ionic strength and pH. When this method was applied to a series of livers from guinea-pigs killed at different time intervals after injection of labelled amino acids, it was found that the protein of one microsomal subfraction had a much higher turnover rate than that of any other (Simkin & Work, 1957a). This fraction was shown by analysis to be a ribonucleoprotein with a protein/nucleic acid ratio around 3:1. Surprisingly enough, a second ribonucleoprotein fraction of similar analytical composition, but different solubility, contained proteins of low turnover rate. Littlefield, Keller, Gross & Zamecnik (1955) fractionated microsomal material from rat liver by an entirely different method and also found a ribonucleoprotein fraction with an especially high turnover rate. Using electron microscopy they were able to identify this fraction as probably identical with the electron-dense granular material attached to the endoplasmic reticulum (Palade & Siekevitz, 1956)—the so-called microsomal particles. Allfrey, Mirsky & Osawa (1957) have found that when isolated thymus nuclei are incubated with labelled amino acids, there is also rapid incorporation into protein, and that in this case the protein fraction of highest specific activity is associated with deoxyribonucleic acid.

Simkin & Work (1957b) have compared the pattern of labelling obtained in cell-free preparations with that observed in intact animals. The protein of fraction C, the ribonucleoprotein fraction which showed a particularly high rate of turnover in intact animals, showed a low rate of uptake of isotope in cell-free systems and that of fraction *B*, which had a low rate of turnover in intact animals, became rapidly labelled *in vitro*. It is certainly tempting to suggest that fraction *C* contains a precursor protein which later provides material for synthesis of complete protein by other parts of the synthetic mechanism. It is not at all clear at present why there should be this difference between labelling pattern *in vivo* and *in vitro*, but it may be a reflexion of the apparent failure of the cell-free system to produce complete protein molecules.

If the microsome fraction is indeed concerned with the synthesis of specific, soluble proteins, then it is reasonable to suppose that such proteins might be found to be bound to microsome material, and could be dissociated from it under appropriate conditions. Simkin (unpublished) labelled liver microsome material by incubation with cell sap and radioactive amino acids in the usual way and then transferred the microsome fraction to fresh cell sap containing no labelled amino acids. In the absence of a substrate for the adenosine-triphosphate generating system there was little change in the specific activity of either the microsomal or the cell-sap proteins, but in the presence of an energy source there was a fall in the specific activity of the microsomal protein and a corresponding increase in the specific activity of the soluble protein.

It is clear from preliminary reports that other laboratories using different biological systems are working along similar lines. Thus, Rabinovitz & Olson (1957) prepared labelled reticulocytes by incubation of cells with radioactive leucine. The cells were then disrupted and the microsome material was isolated and incubated with fresh cell sap. In the presence of adenosine triphosphate there was a fall in the specific activity of the microsomal protein and an increase in the specific activity of the haemoglobin fraction. Fractionation of the soluble proteins indicated that a new red-coloured protein was present with a higher specific radioactivity than authentic haemoglobin. Hendler (1957) has reported somewhat similar experiments. Hen oviduct tissue was labelled by incubation of a cell brei with radioactive amino acids and after homogenization of the tissue a ribonucleoprotein-rich insoluble fraction was isolated. This fraction, incubated with fresh cell sap, released radioactive protein into the soluble fraction. The release was again dependent upon the presence of adenosine triphosphate and the system required in addition guanosine triphosphate, phosphoenolpyruvate and co-enzyme A. It appears then that the release of labelled protein from microsome material requires much the same conditions whether the microsomes have been labelled in intact- or broken-cell systems. It has been suggested (e.g. Hendler, Dalton & Glenner, 1957;

Palade, 1956) that the system of cavities within the membranes of the endoplasmic reticulum may be associated with the secretion of protein formed by the microsome particles, and it is noteworthy that in cells in which there is very active synthesis of protein, such as the pancreas, the endoplasmic reticulum is very highly developed (Palade, 1955).

All these results taken together make it reasonable to hope that detailed comparison of the process of amino acid incorporation into microsome material both in intact animals and in cell-free systems will provide considerable information as to the mechanism of protein biosynthesis. It is clear also that microsomal proteins with a high turnover rate are particularly closely associated with ribonucleic acid (RNA). We are thus forced to inquire: what is the function of RNA in protein synthesis?

II. PARTICIPATION OF RIBONUCLEIC ACID IN PROTEIN SYNTHESIS

Although there is a broad experimental basis for the belief that RNA plays a special role in protein synthesis, the bulk of the evidence is indirect and some of it is indeed self-contradictory. It is clear that depletion of cellular RNA by exposure of cells to the action of ribonuclease destroys the capacity for protein synthesis (Brachet, 1955), but cell metabolism is such a closely integrated whole that removal of any other major component would be likely to have a similar effect. Ribonuclease treatment also destroys the capacity of cytoplasmic subcellular particles to incorporate amino acids (Allfrey et al. 1953; Zamecnik & Keller, 1954; Littlefield & Keller, 1957), but again, the action of the enzyme is sufficient to destroy organization. Allfrey et al. (1957) have similarly shown that exposure of isolated nuclei to the action of deoxyribonuclease reduces their capacity to incorporate amino acids into protein. In some cases restoration of function has been achieved by incubation of nuclease-treated preparations with the appropriate nucleic acid. In isolated deoxyribonuclease-treated calf-thymus nuclei, restoration can be produced not only by specific deoxyribonucleic acid (DNA), but also by denatured DNA or by enzymic hydrolysates of either DNA or RNA (Allfrey et al. 1957). In other cases restoration of function has been produced by incubation with a mixture of purines and pyrimidines (e.g. Gale & Folkes, 1955; Webster & Johnson, 1955). In the case of isolated liver mitochondria, which incorporate amino acids at about one-fifth the rate of microsomes, Simpson, McLean, Cohn & Brandt (1957) have reported that incorporation was stimulated by exposure to ribonuclease.

Much evidence has accumulated also that protein synthesis in the form of induced enzyme formation by bacteria requires the simultaneous presence of the four nucleotide components of RNA and is accompanied by RNA synthesis (Spiegelman, 1956). Allfrey *et al.* (1957) have suggested that synthesis of protein in isolated thymus nuclei requires the synthesis of RNA. We thought it worth while, therefore, to compare the rate of RNA turnover in different ribonucleoprotein subfractions derived from liver microsomes. Turnover was measured in normal guinea-pigs killed at suitable intervals after injection of ^{32}P-phosphate. It was found that subfraction C, which has a high rate of protein turnover, had a low rate of RNA turnover and that in fraction B, which has a low rate of protein turnover, the rate of RNA turnover was higher than in C (Bhargava, Simkin & Work, 1958). This result appears at first sight to be contradictory to the results obtained using induced enzyme formation as a criterion of protein synthesis, but it is in line with other evidence obtained using mammalian cells (Shigeura & Chargaff, 1957; Clark, Naismith & Munro, 1957). Thus not all of the available evidence supports the concept that synthesis of RNA is associated with protein synthesis. As we have already suggested (Simkin & Work, 1957c), some of the data obtained can be explained more readily by the hypothesis that common intermediates, perhaps nucleotide-amino acid complexes, are involved in the synthesis of RNA and protein (cf. also Crick, this Symposium; Hotchkiss, 1956; Yčas & Brawerman, 1957).

In addition to the above type of evidence, which is somewhat indirect, there is the observation of Hoagland *et al.* (1957), already mentioned, that a soluble amino acid-RNA derivative is formed during the transfer of amino acids to the microsomal particles. This is, as far as we know, the first direct demonstration of participation of RNA during the incorporation of amino acids into proteins, but since synthesis of specific proteins in a cell-free system has not been demonstrated the significance of the observation cannot be fully assessed. This function of SRNA in protein synthesis would appear to be different from the role of RNA in microsome particles. The nature of its function in the latter is still a matter of speculation (see Crick, this Symposium). The possibility that RNA may play more than one role in protein synthesis has often not been given enough consideration. If RNA does have more than one function, this clearly complicates the interpretation of experiments in which changes in overall RNA turnover have been demonstrated, as e.g. during the synthesis of induced enzymes in bacteria.

III. NUCLEIC ACIDS IN THE CONTROL OF PROTEIN SPECIFICITY

There is no doubt that DNA, in the form of bacterial transforming factor, or RNA in the form of virus can enter cells and bring about alterations in the specificity of protein synthesis. The foreign nucleic acid entering a cell is thus able not only to bring about the replication of itself, but also to alter the metabolic potentialities of the cell, providing information for the production of new specific macromolecules of other kinds. Crick (this Symposium) has already outlined the background to the belief that the function of RNA is to act as a template for protein synthesis and, on the basis of this theory, a logically attractive argument can be developed. It is, as he says, difficult to offer any alternative theory which is equally complete and self-consistent, but we must be careful to distinguish logical attractiveness from experimental evidence, and alternative explanations of existing evidence should still be sought.

When either RNA or DNA enters a growing cell it enters a highly organized environment which must have a favourable genetic constitution before any change can be induced. Support for this assertion is provided by the recent work of Schaeffer (1957), who demonstrated that both homo- and hetero-specific transforming DNA are able to enter into cells of *Haemophilus influenzae* to comparable extents. However, only the penetration of the homo-specific DNA results in a high frequency of transformation. If the function of the entering nucleic acid were simply to provide a suitable surface or template for the formation of new protein, it is difficult to see why the genetic limitation should exist. One might rather expect that all virus nucleic acids would be infective towards all types of cell. The fact that this is not so suggests that the nucleic acid influences the metabolic capacities of the cell by interacting with the components of the cell rather than by acting in an independent manner. The existence of genetic limitations would seem to imply that before RNA or DNA can induce transformation in the synthetic mechanism it may have to be so close in structure to an existing cell component that the cell fails to 'recognize' it as a foreign molecule. In this case the new structure would be reproduced by the same mechanism which the cell had already developed for replication of other similar molecules.

Kornberg (1957) has obtained from *Escherichia coli* an enzyme which will catalyse the formation of polynucleotide from the four appropriate deoxyribonucleoside triphosphates. The reaction requires the presence of all four nucleotides, and the enzyme acts only in the presence of a DNA primer, though the latter may be of animal, bacterial or viral origin. This reaction has, therefore, the characteristics which one might expect of a process of self-replication. It is difficult to see why all four nucleotides should be required unless a tetranucleotide structure is being produced and it is difficult to see why primer should be necessary unless the new tetranucleotide structure is being formed on the surface of the old by a process akin to crystallization. It has not yet been established whether this model of DNA synthesis represents the replication of DNA in the intact cell, or whether it is valid for the replication of other kinds of macromolecule. On a more speculative level, Penrose & Penrose (1957) have suggested that self-replication can be achieved with surprisingly simple models; an elaborate structural configuration does not appear to be obligatory.

These speculations are presented, not because they are particularly preferred to other current speculations but rather to emphasize that various alternative mechanisms for replication should be considered and because it is our firm opinion that there is as yet insufficient evidence upon which to base a preference for one mechanism rather than another. Any such preference, as Crick emphasized, partakes of the character of dogma.

REFERENCES

ALLFREY, V. G., DALY, M. M. & MIRSKY, A. E. (1953). *J. Gen. Physiol.* **37**, 157.

ALLFREY, V. G., MIRSKY, A. E. & OSAWA, S. (1957). *J. Gen. Physiol.* **40**, 451.

ASKONAS, B. A., CAMPBELL, P. N., GODIN, C. & WORK, T. S. (1955). *Biochem. J.* **61**, 105.

ASKONAS, B. A., SIMKIN, J. L. & WORK, T. S. (1957). *Symp. Biochem. Soc.* **14**, 32.

BHARGAVA, P. M., SIMKIN, J. L. & WORK, T. S. (1958). *Biochem. J.* (in the Press).

BORSOOK, H., DEASY, C. L., HAAGEN-SMIT, A. J., KEIGHLEY, G. & LOWY, P. H. (1950). *J. Biol. Chem.* **187**, 839.

BORSOOK, H., DEASY, C. L., HAAGEN-SMIT, A. J., KEIGHLEY, G. & LOWY, P. H. (1952). *J. Biol. Chem.* **196**, 669.

BRACHET, J. (1955). *The Nucleic Acids*, vol. 2, p. 475. Ed. by Chargaff, E. & Davidson, J. N. New York: Academic Press.

CLARK, C. M., NAISMITH, D. J. & MUNRO, H. N. (1957). *Biochim. Biophys. Acta*, **23**, 587.

COLVIN, J. R., SMITH, D. B. & COOK, W. H. (1954). *Chem. Rev.* **54**, 687.

CONNELL, G. E. & WATSON, R. W. (1957). *Biochim. Biophys. Acta*, **24**, 226.

GALE, E. F. & FOLKES, J. P. (1955). *Biochem. J.* **59**, 675.

GODIN, C. & WORK, T. S. (1956). *Biochem. J.* **63**, 69.

HEIMBERG, M. & VELICK, S. F. (1954). *J. Biol. Chem.* **208**, 725.

HENDLER, R. W. (1957). *Fed. Proc.* **16**, 194.

HENDLER, R. W., DALTON, A. J. & GLENNER, G. G. (1957). *J. Biophys. Biochem. Cytol.* **3**, 325.

HOAGLAND, M. B., ZAMECNIK, P. C. & STEPHENSON, M. L. (1957). *Biochim. Biophys. Acta*, **24**, 215.

HOTCHKISS, R. D. (1956). *Arch. Biochem. Biophys.* **65**, 302.

HULTIN, T. (1950). *Exp. Cell Res.* **1**, 376.

KORNBERG, A. (1957). *Advanc. Enzymol.* **18**, 191.

LINDEGREN, C. C. (1955). *Nature, Lond.* **176**, 1244.

LITTLEFIELD, J. W. & KELLER, E. B. (1957). *J. Biol. Chem.* **224**, 13.

LITTLEFIELD, J. W., KELLER, E. B., GROSS, J. & ZAMECNIK, P. C. (1955). *J. Biol. Chem.* **217**, 111.

MEINHART, J. O. & SIMMONDS, S. (1955). *J. Biol. Chem.* **216**, 51.

PALADE, G. E. (1955). *J. Biophys. Biochem. Cytol.* **1**, 59.

PALADE, G. E. (1956). *J. Biophys. Biochem. Cytol.* **2**, 417.

PALADE, G. E. & SIEKEVITZ, P. (1956). *J. Biophys. Biochem. Cytol.* **2**, 171.

PENROSE, L. S. & PENROSE, R. (1957). *Nature, Lond.* **179**, 1183.

RABINOVITZ, M. & OLSON, M. E. (1957). *Fed. Proc.* **16**, 235.

SCHAEFFER, P. (1957). *C.R. Acad. Sci., Paris*, **245**, 375.

SHARON, N. & LIPMANN, F. (1957). *Arch. Biochem. Biophys.* **69**, 219.

SHIGEURA, H. T. & CHARGAFF, E. (1957). *Biochim. Biophys. Acta*, **24**, 450.

SIEKEVITZ, P. (1952). *J. Biol. Chem.* **195**, 549.

SIMKIN, J. L. (1955). *Commun. 3rd Int. Congr. Biochem.*, Brussels, p. 74.

SIMKIN, J. L. & WORK, T. S. (1957a). *Biochem. J.* **65**, 307.

SIMKIN, J. L. & WORK, T. S. (1957b). *Biochem. J.* (in the Press).

SIMKIN, J. L. & WORK, T. S. (1957c). *Nature, Lond.* **179**, 1214.

SIMPSON, M. V. (1955). *J. Biol. Chem.* **216**, 179.

SIMPSON, M. V., McLEAN, J. R., COHN, G. L. & BRANDT, I. K. (1957). *Fed. Proc.* **16**, 249.

SIMPSON, M. V. & VELICK, S. F. (1954). *J. Biol. Chem.* **208**, 61.

SPIEGELMAN, S. (1956). *Enzymes: Units of Biological Structure and Function*, p. 67. Ed. Gaebler, O. H. New York: Academic Press.

STEINBERG, D. & ANFINSEN, C. B. (1952). *J. Biol. Chem.* **199**, 25.

STEINBERG, D. & MIHALYI, E. (1957). *Annu. Rev. Biochem.* **26**, 373.

STEINBERG, D., VAUGHAN, M. & ANFINSEN, C. B. (1956). *Science*, **124**, 389.

TURBA, F. & ESSER, H. (1955). *Biochem. Z.* **327**, 93.

VAUGHAN, M. & ANFINSEN, C. B. (1954). *J. Biol. Chem.* **211**, 367.

WEBSTER, G. C. & JOHNSON, M. P. (1955). *J. Biol. Chem.* **217**, 641.

YČAS, M. & BRAWERMAN, G. (1957). *Arch. Biochem. Biophys.* **68**, 118.

ZAMECNIK, P. C. & KELLER, E. B. (1954). *J. Biol. Chem.* **209**, 337.

ZAMECNIK, P. C., KELLER, E. B., LITTLEFIELD, J. W., HOAGLAND, M. B. & LOFTFIELD, R. B. (1956). *J. Cell. Comp. Physiol.* **47**, suppl. 1, 81.

FORMATION OF AMYLASE IN THE PANCREAS

By F. B. STRAUB

Institute of Medical Chemistry, University of Budapest, Hungary

During the last four years we have studied the formation of amylase under a variety of *in vitro* conditions in the hope of obtaining information about the mechanism of protein biosynthesis. The main results of our findings strengthen the view that intermediate stages in protein biosynthesis do exist. I shall try to convince you that the study of these intermediate forms may help in understanding the detailed mechanism of the formation of protein molecules.

The increase of amylase activity in pancreas tissue slices *in vitro*, under aerobic conditions in the presence of amino acids, was first described by Hokin (1951 *a*, *b*) and has been confirmed by several authors. The dependence of the phenomenon on oxidative phosphorylation and on the presence of amino acids suggested that it represents a real synthesis of protein molecules, although the incorporation of radioactive amino acids into the amylase molecule has not yet been demonstrated.

Next, it was found by Hessin (1953) that some increase of amylase activity can be detected in a sucrose homogenate of the pancreas in the presence of ATP, and that this activity is bound to the heavier granules of the cells. We have found (Ullmann & Straub, 1954, 1955, 1956, 1957 *a*) that if a high concentration of ATP is present, the increase of amylase activity in a homogenate or in the fraction of heavier granules ('mitochondria'), is of the same order of magnitude as that found in the tissue slice. The rate of amylase formation in this case is the same as that occurring *in vivo*. We found later that an extract of acetone-dried pancreas behaves in a similar way. In the presence of ATP and amino acids, a considerable increase of amylase activity occurs (Ullmann & Straub, 1957 *a*).

In order to establish that the increase of amylase activity in cell-free systems is due to *synthesis* of amylase, we have investigated the effect of several substances known to inhibit protein synthesis. Chloramphenicol, *p*-fluorophenylalanine, diaminoacridines and ribonuclease were all found to abolish the increase of amylase activity. In the soluble system (extract of acetone-dried pancreas), where permeability barriers no longer exist, these substances act in extremely low concentrations, which indicates their action to be highly specific. It could thus be safely concluded that amylase is formed through a synthetic reaction in a cell-free system.

A deeper understanding of these phenomena was reached when we were able to measure the incorporation of radioactive amino acids into the amylase molecule. We have used a micro-scale isolation method which has enabled us to isolate the amylase contained in a few mg. of pancreas in a practically pure state. The method is rather simple: amylase is adsorbed by Lintner starch at 0° and can then be washed exhaustively until free of other proteins. The starch-amylase complex is then decomposed by 10 min. incubation at 37° C., leaving practically pure amylase in solution, with a yield of 30–60 %.

Using this method it was found that amylase is extremely rapidly labelled in an aerobically incubated tissue slice. No incorporation was found in a homogenate or in a soluble system. However, there is a similar increase of amylase activity in the first 30–60 min. in both cases. We have assumed that amylase is formed in the pancreas in at least two distinct steps, a protein precursor being first formed and later synthetically transformed into the enzymatically active amylase molecule. In the cell-free systems the second step only occurs, the synthetic formation of amylase from the preformed precursor, whereas the intact cells of the tissue slices are able to synthesize the entire molecule *de novo*.

The concept of protein precursors, e.g. Northrop's 'urprotein' has received little attention lately, owing to the undoubted success of the *de novo* theory. The spectacular experiments of Hogness *et al.* (1955) and Speigelmann, Halvorson & Ben-Ishai (1955), the very careful experiments of Work's group (Askonas, Campbell, Godin & Work, 1955), those of Simpson and Velick (1954), and Simpson (1955) and others have failed to find any evidence for the existence of intermediate steps in protein biosynthesis *in vivo*. The results of Steinberg & Anfinsen (1952), who found unequal labelling of the same amino acid in different parts of the protein molecule, were explained away as artifacts of the *in vitro* conditions.

Another line of evidence suggesting intermediate stages in protein synthesis was started by Peters (1953), who observed a considerable time-lag in the labelling of serum-albumin in liver slices *in vitro*. Green & Anker (1955) have found a similar lag *in vivo*. Junqueira, Hirsch & Rothschild (1955) observed an even longer lag-period in the appearance of the label in the protein of the pancreatic secretion.

These results were rightly criticized from two opposing directions. The *in vivo* experiments are open to the objection that the time-lag is due to transport and secretion processes. The *in vitro* experiments of Peters were criticized as not reflecting the *in vivo* conditions, or, alternatively, because he investigated the soluble serum-albumin appearing in the supernatant of a homogenate, and in this way could not exclude the possible role of transport processes.

Short of isolating a precursor protein which can be transformed into the final product, the next best argument for the existence of such a precursor is still the evidence of a time-lag in the labelling of the protein. Garzó, T.-Szabó & Straub (1957) have investigated the time-course of labelling of the amylase molecule in aerobically incubated tissue slices of pigeon pancreas, taking care to isolate the amylase after complete destruction of the cell structure.

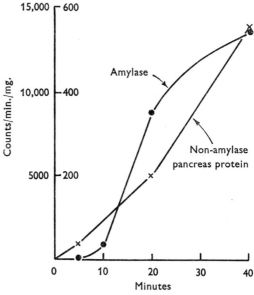

Fig. 1. Labelling of amylase and other proteins in pigeon pancreas tissue slices. The outer numbers of the ordinate refer to the labelling of amylase, the inner numbers to the labelling of the non-amylase proteins.

One or two pigeon pancreases were sliced, 200–300 mg. were placed in 9 ml. Krebs-bicarbonate saline solution in presence of 0·2 % casein hydrolysate and 10^6 counts per minute glycine-1-^{14}C. The vessels were flushed with O_2—CO_2 mixture and then shaken at 37° in a water bath for various periods of time. After incubation, the slices were removed, homogenized with acetone and then dried. From the dry powder amylase was extracted with a phosphate-saline solution. The residue and the amylase isolated from the extract were separately used for determination of radioactivity. The result of such an experiment is shown in Fig. 1. It is evident that the labelling of amylase shows a time-lag of about 10 min., although the increase of amylase activity is linear. On the other hand it will be remarked that the rate of labelling after 10 min. is twenty times faster in the amylase than in the other proteins.

In another type of experiment T.-Szabó & Garzó (1957) studied the intracellular localization of these processes. The microsome ribonucleoprotein fraction of liver cells is known to be labelled first among the cellular components. However, not much is known about the nature of these microsome proteins. We tried to approach this question with the use of our micro-scale amylase isolation method. For reasons of simplicity pigeon pancreas tissue slices were labelled *in vitro*, homogenized and fractionated after different

Table 1. *Incorporation into the total proteins of pancreas fractions**

No. of experiments	Time of incubation (min.)	Glycine-1-^{14}C added (counts per min. × 10^6)	Nuclei	Zymogen granules	Mitochondria	Microsomes I	Soluble fraction I	Microsomes II	Soluble fraction II
				Specific activity of subcellular fractions of pigeon pancreas (counts/min./mg. protein)					
1	60	4·5	1850	7795		3790			
2	60	4·5	1350	5270		2860	2800		
3	60	2·2	855	4370		2340	945		
4	60	3·0	1280	1780	2100	1330	1100		
5	30	3·6	1700	4700	4280	3300	3000		
	60	3·6	1100	3500	5400	3500	7900		
	120	3·6	5200	13600	13100	4700	17000		
6	30	3·6	1200	3600	3700	2580	3100		
	60	3·6	2350	3700	6000	4270	9450		
	120	3·6	4460	9600	12900	9000	25000		
7	60	2·9	782	3960	2720	2940	3570	2500	2260
8	60	3·6	3200	8000	—	4950	19300	10700	6450
9	60	2·5	736	3140	1380	1700	1920	5500	3160
10	30	2·5	1520	4560	—	—		4025	3170
	120	2·5	6970	17500	—	—		13050	10900

* Microsomes were removed by two different methods: (1) by precipitating them with an acetate buffer at pH 5·4 (microsomes I, leaving the soluble fraction I in solution); (2) by centrifuging at neutral pH at 100,000 g (the precipitate is denoted microsomes II, its supernatant being the soluble fraction II).

periods of incubation. The nuclear, zymogen, mitochondrial, microsome and the plasma fractions were separated, washed with appropriate solvents and then analysed for the radioactivity of the total protein they contain.

The time-course of labelling of the total proteins of these fractions is different from that observed in liver cells. Although we were able to confirm the rapid labelling of the microsome fraction of rat liver preparations, we have consistently found that of the mitochondrial fraction of the pancreas to run ahead of the microsomes (see Table 1). The discrepancy is explained, however, by the following experiments. We have labelled the tissue slice, homogenized the tissue in sucrose after incubation, and isolated the amylase from each fraction after drying the separated fractions with acetone. Actually the nuclear fraction, after washing with a citrate solution, contained no amylase, so that it could not be investigated. Moreover, the plasma fraction, i.e. the fraction remaining in solution after removal of all the granular fractions, is very likely a waste-bin of the amylase lysed from different fractions and is therefore not relevant (Table 2). Fig. 3 shows the results of such experiments, in which the radioactivity of the amylase in the zymogen, mitochondrial and microsome fraction were measured at different times. Two interesting facts emerge. At 10 min., the radioactivity (counts per mg. protein) of the microsome-amylase is the highest. On the other hand, the radioactivity increases more or less linearly in the amylase of the microsome fraction, whereas it shows a pronounced time-lag in the

Table 2. *Incorporation into the amylase of pancreas fractions*

Subcellular fractions		Time of incubation (min.)	Purity of amylase	Specific activity (counts/min./mg. protein)
Zymogen	1	30	0·87	1020
granules	2	30	0·86	705
	3	15	0·73	595
		30	—	—
	4	10	0·73	745
		30	0·80	3840
Mitochondria	1	30	0·87	1020
	2	30	1·03	7450
	3	15	0·80	1720
		30	1·08	10600
	4	10	1·15	1000
		30	1·00	11000
Microsomes	1	30	0·57	3900
	2	30	0·64	8820
	3	15	0·39	1960
		30	0·34	7100
	4	10	0·16	1200
		30	0·24	5800
Soluble	1	30	1·00	438
fraction	2	30	0·92	1210
	3	15	0·91	1310
		30	0·77	4700
	4	10	1·00	1025
		30	1·04	5850

mitochondrial amylase. The labelling of the zymogen granule amylase always lags behind these two and is intermediate with regard to its time-course. Taken together with our former results, we may say that the precursor protein of amylase is formed in the microsomes (the precursor itself is not isolated by our method of amylase isolation). Some of the precursor is transformed into amylase already in the microsome fraction, but the greater part of it is transferred into the mitochondrial fraction, where it is transformed in bulk into amylase. After a considerably longer time the amylase is transferred into the zymogen granules. The transfer of the precursor from microsomes to mitochondrial fraction explains why the latter fraction was found by us to be already highly labelled after only 10 min. of incubation.

I have to guard myself against criticisms by cytologists, by stating that I do not know whether the transformation of precursor into amylase occurs in the classical mitochondria themselves. I refer only to the 'mitochondrial' fraction, which is sedimented from a sucrose solution between 1000 and 14000 *g*. Alternatively it is conceivable that enzyme synthesis starts in microsomes attached to the endoplasmic reticulum, and when enough precursor is synthesized, a vesicle is formed *in vivo* (or only during homogenization) which sediments together with the mitochondria. It is even conceivable that such vesicles themselves are later transformed into secretory granules.

Before leaving the subject of localization of amylase synthesis an interesting observation should be mentioned. When discussing the experiment shown in Table 2, it was stated that the labelling of the purified amylase from microsomes is the highest, calculated per mg. protein of the purified amylase. Actually our micro-scale amylase-isolation method yields practically pure amylase from all fractions, except the microsomes. The amylase isolated from the microsomes is never pure, it has usually 30–60 % of the maximal enzyme activity per mg. protein. Also this microsomal amylase always contains considerable quantities of pentose. Although the potentialities of this finding are evident, at present it is rather a nuisance. We have some evidence that all of the radioactivity of this fraction belongs to the amylase it contains—and that means that its labelling is even higher than is shown in Table 2—but this requires more rigid proof.

I should like to return now to our other line of investigation. A precursor is first synthesized *de novo* in the intact cell, and then transformed synthetically into the complete amylase molecule. The latter step is observed also in cell-free homogenates. Such a transformation also goes on in the soluble system described. What other factors are needed? I have already mentioned that RNA is implicated in the process, as ribonuclease prevents the increase of amylase activity. ATP is needed, which shows the synthetic nature of the reaction. We have also found that amino acids are needed, and that the reaction is inhibited by amino acid analogues. But, on the other hand, no incorporation of radioactive glycine occurs. This discrepancy has prompted us to investigate the possibility that not all of the amino acids may be necessary. It turned out that this is indeed the case. In a soluble system prepared from pigeon pancreas, Ullmann & Straub (1957*b*) found that only two amino acids are needed—threonine and arginine. There is practically no increase of amylase activity if no amino acid is added (Fig. 2); some increase if threonine or arginine alone is present; and a maximal increase if both are added at a level of 0·5 mg./ml. As there is always some amino acid present due to autolysis, which may explain why threonine alone already has considerable activity, we may conclude that these two amino acids are both necessary. We have ascertained that neither of them is built into the amylase molecule, as there was definitely no incorporation of radioactive amino acid into the amylase, when the soluble system contained a hydrolysate of labelled algal protein containing active threonine and arginine. It is worth mentioning that the soluble system from dog pancreas requires histidine and threonine, but not arginine. I am

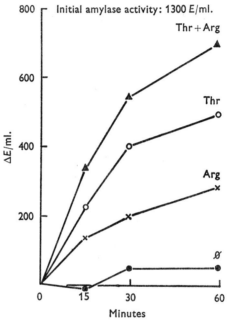

Fig. 2. Increase of amylase activity in a soluble system from pigeon pancreas. $0 \cdot 016$ M-ATP and Krebs-Ringer present in all cases. Addition of amino acids: Thr: D-, L-threonine $1 \cdot 0$ mg./ml.; Arg: L-arginine 1 mg./ml.; Thr + Arg: $0 \cdot 5$ mg. of each.

unable to offer any explanation at present as to what these two amino acids are needed for if they are not built into the protein. However, the next observation gives some support to the view that these amino acids are activated during the process.

Ullmann has tried to fractionate the soluble system, which shows some increase of amylase activity. The soluble system is after all nothing but a water extract of the acetone-dried tissue. From such a water extract acetate buffer precipitates a fraction at about pH 5, which is sometimes active and sometimes inactive in respect of amylase formation. We found that the supernatant contains an accessory factor, which by itself is completely inactive, but which, when added to the acetate precipitate, gives a fully active system (Fig. 3) This factor may be replaced by a similar fraction prepared from acetone-dried liver extract and is therefore not specific for the formation of amylase. We were also able to replace this factor in experiments with dog pancreas by a purified enzyme preparation, which is considerably enriched in the acyl-activating enzyme specific for threonine and histidine.

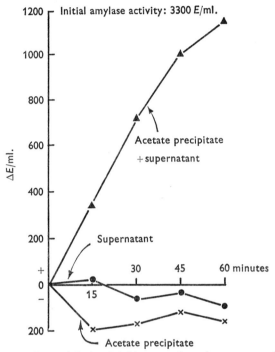

Fig. 3. Increase of amylase activity in a soluble system from dog pancreas. Acetone powder extracted with 5 volumes water, 2 ml. of this extract + 1 ml. water + 0·3 ml. 0·2 M-acetate buffer pH 4·2 centrifuged at 0° yields the acetate precipitate. The supernatant of this is dialysed for 24 hr. at 0°. 0·2 ml. of the supernatant and all of the acetate precipitate is used in each experiment in a total volume of 1 ml., supplemented with ATP, threonine, histidine and Krebs-Ringer.

The acetate precipitate of the soluble system is rather labile and it is seriously damaged by treatment with ribonuclease. If it is partly or nearly completely inactivated, e.g. by preparing it from a stored acetone powder, the activity can be considerably increased by addition of pancreas micro-some preparations, which are themselves completely inactive in bringing about the formation of amylase (Fig. 4). It appears, therefore, that the transformation of the precursor protein into amylase requires an array of factors, including perhaps microsomal RNA and acyl-activating enzymes.

Summing up, our experimental evidence supports the existence of inter-mediary stages in enzyme formation. It represents a new approach in the sense that we measure the formation of one defined protein molecule.

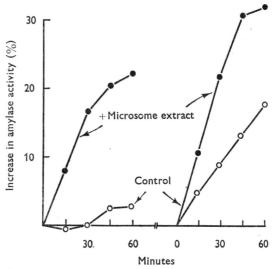

Fig. 4. Effect of microsome extract on the increase of amylase activity in the soluble system from dog pancreas. Left: aged soluble system. Right: freshly prepared soluble system. Microsome extract: particles were removed from a sucrose-homogenate of dog pancreas at 14000 g and then microsomes were precipitated at pH 5·4. The precipitate was dried with acetone and the powder extracted with 10 volumes of water. 0·1 ml. of this extract was used in 1 ml. total volume of each experiment. Control: complete soluble system without microsome extract.

REFERENCES

Askonas, B. A., Campbell, P. N., Godin, C. & Work, T. S. (1955). *Biochem. J.* **61**, 105.

Garzó, T., T.-Szabó, M. & Straub, F. B. (1957). *Acta Physiol. Hung.* **12**, 299.

Green, H. & Anker, H. S. (1955). *J. Gen. Physiol.* **38**, 283.

Hessin, R. B. (1953). *Biochimiya*, **18**, 462.

Hogness, D. S., Cohn, M. & Monod, J. (1955). *Biochim. Biophys. Acta*, **16**, 99.

Hokin, L. E. (1951 a). *Biochem. J.* **48**, 320.

Hokin, L. E. (1951 b). *Biochem. J.* **50**, 216.

Junqueira, L. C. U., Hirsch, C. G. & Rothschild, H. A. (1955). *Biochem. J.* **61**, 275.

Peters, T. (1953). *J. Biol. Chem.* **200**, 461.

Simpson, M. V. (1955). *J. Biol. Chem.* **216**, 179.

Simpson, M. V. & Velick, S. F. (1954). *J. Biol. Chem.* **208**, 61.

Spiegelmann, S., Halvorson, H. O. & Ben-Ishai, R. (1955). *A Symposium on Amino Acid Metabolism*, p. 124. Ed. McElroy, W. D. & Glass, B. Baltimore: Johns Hopkins Press.

Steinberg, D. & Anfinsen, C. B. (1952). *J. Biol. Chem.* **199**, 25.

Straub, F. B. (1957). *Acta Physiol. Hung.* **12**, 295.

T.-Szabó, M. & Garzó, T. (1957). *Acta Physiol. Hung.* **12**, 303.

Ullmann, A. & Straub, F. B. (1954). *Acta Physiol. Hung.* **6**, 377.

Ullmann, A. & Straub, F. B. (1955). *Acta Physiol. Hung.* **8**, 279.

Ullmann, A. & Straub, F. B. (1956). *Acta Physiol. Hung.* **10**, 137.

Ullmann, A. & Straub, F. B. (1957a). *Acta Physiol. Hung.* **11**, 11.

Ullmann, A. & Straub, F. B. (1957b). *Acta Physiol. Hung.* **11**, 31.

THE BIOSYNTHESIS OF OLIGO- AND POLYSACCHARIDES

By M. STACEY, F.R.S.

Chemistry Department, The University, Birmingham 15

There has been intense activity and dramatic progress in this field during the past decade or more, and many comprehensive reviews are available for study (Hassid & Doudoroff, 1950; Hehre, 1951; Barker & Bourne, 1953; Stacey, 1954). In this account some of the more salient points will be put forward and an effort will be made to bring the subject up to date.

Tribute must be paid to the outstanding pioneer work of Cori, Swanson & Cori (1945) and of C. S. Hanes (1940 a, b) on phosphorylase, which pointed out the direction for further investigations. Now a variety of water-soluble enzymes has been isolated and these have been used to synthesize many types of polysaccharide. In recent weeks the enzymatic synthesis of bacterial cellulose has been claimed and this, if substantiated, will prove to be of outstanding importance. Commercially the synthesis of the polysaccharide, dextran, by isolated and purified enzymes has been shown to be feasible.

My colleagues Barker & Bourne (1953) neatly summarized the biosynthesis situation by stating that a 'master pattern' had now emerged as in the following equation:

$$Gt—O—X + H—O—Gr \rightleftharpoons Gt—O—Gr + X—O—H$$

where Gt is a glycosyl-residue, X is the aglycone part and H—O—Gr is a receptor molecule on which the polysaccharide is built. Each step in the polymerization process involves the transference of the group Gt from —OX to —O—Gr giving what is now called a 'transglycosylation'. This involves the breaking of single glycosidic linkages and reforming them into different types of glycosidic linkages which may perhaps now be termed 'polymeric' linkages.

It was thought until quite recently that three components only were necessary for the biosynthesis to proceed, namely, the 'glycosidic monomer' (substrate), the enzyme system, and the 'primer' or specific receptor molecule. As part of the mechanism it must be envisaged that enzyme-glycosidic linkages are involved in the intermediate transfer stage, i.e. that the enzyme first goes on to Gt then off as in a shuttle system, thus

$$Gt—O—X + enzyme \rightarrow Gt—E + HOX$$
$$Gt—E + H—O—Gr \rightarrow Gt—O—Gr + enzyme$$

In many systems, however, the glycosidic transfer reactions, as demonstrated in a striking way by Leloir and his school, are brought about by uridine diphosphate glucose (UDPG) as an intermediate glycosidic substrate (Leloir & Cardini, 1955). Thus enzyme preparations from wheat, corn, pea and bean germs can catalyse the reversible formation of sucrose from UDPG and fructose. Moreover by replacing the fructose by other ketoses in the system various related glucose-ketose disaccharides can be synthesized. Again by replacing the glucose in UDPG by another hexose such as glucosamine the latter can be transferred from it to suitable receptor molecules. A polymer of glucosamine resembling chitin has recently been synthesized in this way, and claims have been made for the use of UDPG as the substrate in the synthesis of cellulose by a *Bacterium xylinum* enzyme (Glaser, 1957; Greathouse, 1957).

Whether or not UDPG is a common component of all synthesizing enzyme systems, acting perhaps as a 'leveller' of the energy balance considerations, remains to be seen. It does appear, however, that in the natural state the simple glycosides form sharply specific substrates for a wide variety of enzyme systems. Thus α-D-glucose-1-phosphate is sharply specific for polysaccharide formation with most phosphorylases and sucrose (α-glucose-$(1 \rightarrow 2)$-βO-fructofuranose) is specific for dextran formation by many dextransucrases and for levan formation by levansucrases.

The phosphorylases have been obtained crystalline in a high state of purity, while dextransucrases also have been highly purified. As expected the enzymes are proteins.

It has been of some considerable interest to find that the same or closely related complex polysaccharides can be synthesized by a variety of enzymes acting upon different glycosidic substrates. Thus the straight chain component of starch, amylose (blue-staining with iodine) normally made from glucose-1-phosphate by phosphorylases from many plant and microbial sources, can also be made from maltose (α-glucose $(1 \rightarrow 4)$ glucose) by acting upon it with an enzyme from *Escherichia coli* (termed amylomaltase) if at the same time one removes the liberated glucose units by notatin oxidase. Thus

$$n \text{ maltose} \rightarrow \underset{(amylose)}{(\text{glucose})_n} + n \text{ glucose} \rightarrow (\text{gluconic acid})$$

Straight chain dextrans which consist largely of α-$(1 \rightarrow 6)$ linked glucose units and are normally made from sucrose by bacterial enzymes, can also be made from glucose-1-phosphate and from dextrins by suitable enzymes.

Studies on the biosynthesis of the highly branched amylopectin component of starch opened up an exciting new field of study and indeed pointed the way to the possibility of 'tailor-making' to size and shape by enzymic means a polysaccharide macromolecule. In 1944 Haworth, Peat & Bourne succeeded in isolating from potato extract the so-called 'Q-enzyme' or branching enzyme which with phosphorylase (or P-enzyme) would act upon α-D-glucose-1-phosphate to produce the highly branched polysaccharide component of starch, amylopectin. Later it was shown that Q-enzyme, in the absence of inorganic phosphate, would act upon the long straight chain molecule of amylose and convert it into amylopectin. The mode of action of this enzyme was clearly that of a 'break and make type', i.e. the chain of amylose which consists of at least 200 α-$(1 \to 4)$ linked D-glucose units, was broken at every twentieth such link and cross-linked through the $1 \to 6$ positions to form the highly ramified structure of amylopectin. By allowing mixtures of P- and Q-enzymes in different proportions to act upon glucose-1-phosphate it was possible to synthesize polysaccharide mixtures closely resembling certain natural starches. It has also been found possible to act upon amylose from the potato with a Q-enzyme extracted from certain protozoa, e.g. *Polytomella coecae*, to synthesize amylopectin. In a similar way the enzymes responsible for synthesizing dextrans have been shown to act in sequence. Some dextrans possess long chain molecules consisting almost entirely of long glucose chains having α-$(1 \to 6)$ glucose linkages. Other dextrans have cross links of the $(1 \to 4)$ type, the $(1 \to 3)$ type or both. Absence of magnesium ions favours the formation of straight chain dextran.

An interesting field of research was discovered when note was taken of the work of Monod and others, who showed that accumulation of glucose during the action of amylomaltase on maltose prevented the polymerization of glucose beyond about 10 units (Monod & Torriani, 1950; Barker & Bourne, 1952). With various enzyme systems it was found possible to use mono-, di- and oligosaccharides as receptor groups which would compete with the growing polysaccharide chains for the substrate and eventually stop the growth of the macromolecules.

At Birmingham we used the competitive action of different receptor molecules to study some aspects of the transglycosylation reaction in the formation of tri- and oligosaccharides. An account of this has already been given (Stacey, 1957). The work was helped by our development of chromatographic and ionophoretic techniques and particularly by infra-red studies, by which the nature and geometry of glycosidic linkages can readily be decided.

Koepsell *et al.* (1953) showed that oligosaccharides are formed when the dextran sucrase from *Leuconostoc mesenteroides* (NRRL B512) acts on sucrose in the presence of several simple sugars.

In our laboratories we found that when *Betacoccus arabinosaceus* (Birmingham strain) was grown on sucrose with certain added salts it formed a high molecular dextran containing mainly $(1\to6)$-α-linked glucose units and with some $(1\to3)$ and $(1\to4)$ branch points (Bailey, Barker, Bourne & Stacey, 1954, 1955). If to the same medium various simple sugars such as glucose, mannose, etc. were added, growth usually occurred but the molecular size of the polysaccharide was diminished in a remarkable way. With 40% of added glucose no growth at all occurred, with 10% of glucose highly polymeric dextran was formed, while with 20% of glucose dextran production was negligible and instead there was formed a mixture of isomaltose, isomaltotriose and higher homologues. Addition of increasing amounts of maltose had a similar effect. Indeed no dextran was produced, instead there were detected unchanged maltose, panose [O-α-D-glucopyranose $(1\to6)$-O-α-D-glucopyranosyl-$(1\to4)$-D-glucopyranose], higher oligosaccharides, fructose and fructose-containing di- and tri-saccharides. It was further shown by presenting additional sugars as receptors such as methyl glucoside, cellobiose, fructose, etc. that addition to them of glucose took place, with the formation of a wide range of oligo-saccharides. The same effect could be achieved with isolated enzymes and evidence was obtained of the synthesis of a wide variety of linkages.

Previously, panose had been obtained by a dextran sucrase enzyme acting upon a sucrose-maltose mixture while a novel disaccharide, leucrose, (5-O-α-D-glucopyranosyl)-D-fructose, had been obtained by a similar dextransucrase preparation (Stodola, Koepsell & Sharpe, 1952), thus:

glucopyranose $(1\to2)$ fructofuranose + glucose $(1\to4)$-α-glucopyranose $\rightleftharpoons$
   (sucrose)

   glucopyranose $(1\to6)$-α-glucopyranose-$(1\to6)$-glucopyranose + fructose
   (maltose)

With these systems we have been able to study in some detail the actual synthesis of one special type of linkage (Bailey, Barker, Bourne, Stacey & Theander, 1957). It was found possible to use a high proportion of a mono-saccharide receptor molecule, e.g. lactose, with a sucrose substrate, in the presence of growing *Betacoccus* cultures. With lactose which contains glucose and galactose units, it was possible from examination of the resulting trisaccharides formed by addition of a single glucose unit, to decide which part of the molecule was acting as a so-called 'chain-initiator'. After growth of the organism full use was made of charcoal columns to separate the oligosaccharides, which were then examined by the classical and newer methods of carbohydrate chemistry. In this way we synthesized a novel type of branched trisaccharide having the structure O-β-D-galactopyranose-

$(1 \rightarrow 4)$-O-$[\alpha$-D-glucopyranosyl-$(1 \rightarrow 2)$-D-glucopyranose] as shown in Fig. 1.

At a later stage it was found possible to achieve the synthesis using cell-free enzymes which provided the first enzymic synthesis of a branched trisaccharide. The novelty lies in the fact that the reducing $\alpha\beta$-D-glucose unit of the disaccharide was the acceptor of the other glucose unit enzymically transferred from the sucrose. Normal addition previously of hexose units as in amylose synthesis had always apparently taken place at non-reducing ends of chains. The work also revealed a method for the enzymic synthesis of the rare α-$(1 \rightarrow 2)$-glucosidic linkage joining glucose to glucose.

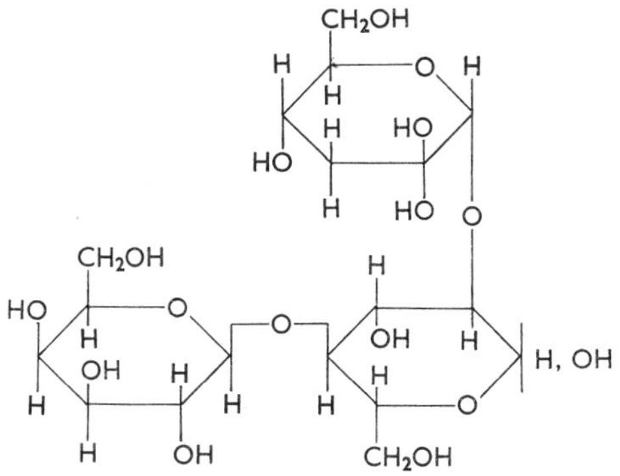

Fig. 1

Previously such a linkage had been detected in a mixture of no less than ten discaccharides made from glucose by *Aspergillus niger* (NRRL 330) (Peat, Whelan & Hinson, 1955). The mechanism for this first synthesis of a branched trisaccharide is probably a general one for branched oligosaccharides and may happen in other cases where the reducing group at the end has the correct structure to act as a receptor. It points to a reasonable route to the direct synthesis of a branched polysaccharide which does not involve the breaking of a linear polymer and remaking it, as in the case of the formation of amylopectin and glycogen.

It might well explain, too, the formation of such complex polymers as the plant gums, where a number of oligosaccharides would first be formed and then cross-linked together to form highly branched macromolecules in a manner similar to the well-known 'block polymers' made chemically. It could offer an explanation for the formation of the highly branched dextrans. As a further example of the synthesis of the $(1 \rightarrow 2)$ type of linkage it was possible to present the same enzyme synthesis with cellobiose as a potential chain initiator (Barker, Bourne, Grant & Stacey, 1956). This did indeed behave in a manner analogous to that of lactose and gave rise to the branched trisaccharide O-β-D-glucopyranosyl-$(1\rightarrow4)$-O-$[\alpha$-D-glucopyranosyl $(1\rightarrow2)]$-D-glucopyranose, the structure of which (Fig. 2) was proved by a variety of methods, including application of the valuable infra-red technique.

It was of great interest to find that both the new branched chain trisaccharides gave rise on enzymic hydrolysis to the same $(1\rightarrow2)$ linked disaccharide, 2-O-α-D-glucopyranosyl-D-glucose. Such a compound was also made by W. J. Whelan (1956) and probably by Japanese workers ('Kojibiose' of Shibasaki & Aso, 1955).

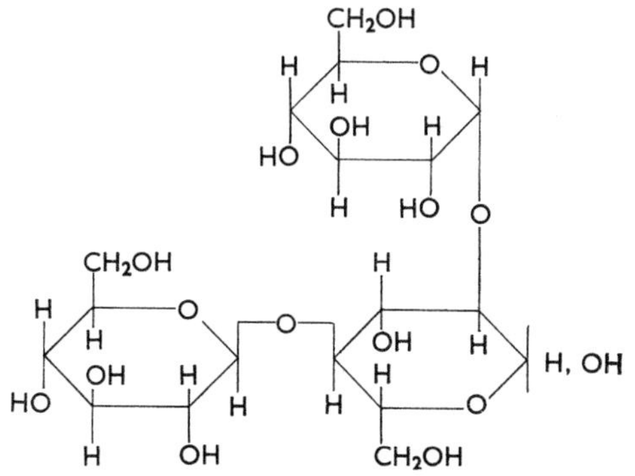

Fig. 2

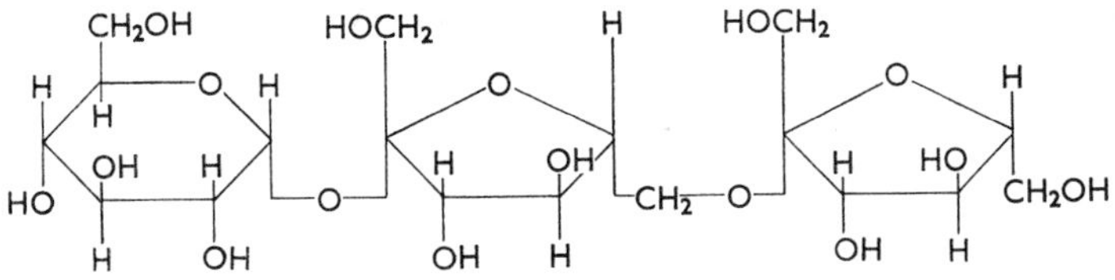

Fig. 3

An interesting trisaccharide was described by Bell and his colleagues (Albon, Bell, Blanchard, Gross & Rundell, 1953). This was made by the action of yeast invertase on concentrated sucrose solutions. It was termed 'kestose' and shown to be O-α-D-glucopyranosyl-$(1\rightarrow2)$-O-β-D-fructonosyl-$(6\rightarrow2)$-β-D-fructofuranoside (Fig. 3). During the preliminary study of the formation by *Aspergillus niger* (strain 152) of the polysaccharide, nigeran, a polyglucosan which contains alternating α-$(1\rightarrow4)$ and α-$(1\rightarrow3)$ linkages, we encountered two non-reducing trisaccharides containing two molecules of fructose and one of glucose (Barker, Bourne & Carrington, 1954). The major trisaccharide was shown to have the structure shown in Fig. 4, i.e. it was O-α-D-glucopyranosyl $(1\rightarrow2)$-O-β-D-fructofuranosyl $(1\rightarrow2)$-β-D-fructofuranoside, and the mechanism of its formation involved a transfructosylase action. The synthesis of oligosaccharides by moulds acting upon sucrose has been demonstrated by several other workers. Thus a commercial mould enzyme preparation, Takadiastase, acting upon sucrose, gave a trisaccharide shown to have the same structure as our trisaccharide (Bealing & Bacon, 1951; Bacon & Bell, 1953).

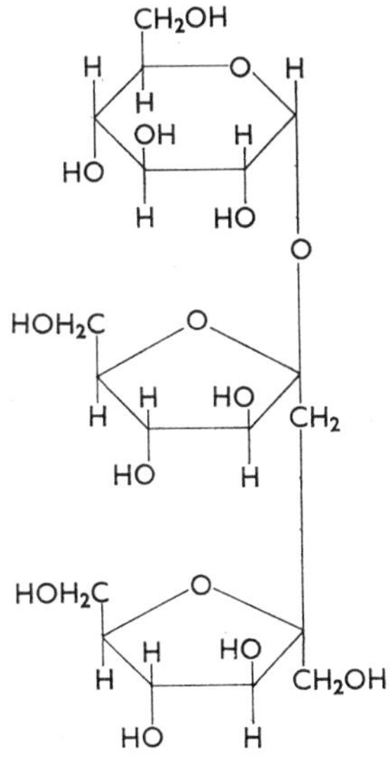

Fig. 4

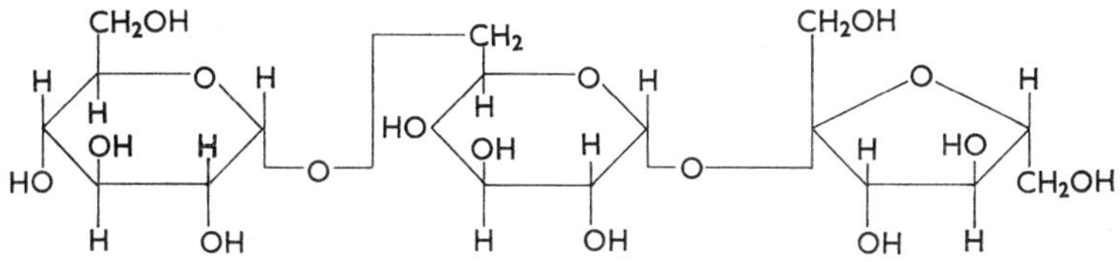

Fig. 5

As part of yet another study of the remarkable synthesizing activities of *Aspergillus niger* (strain 152) we have discovered a new trisaccharide which we consider to be of great importance for studies of oligosaccharide synthesis. A cell-free mould extract was allowed to act upon a mixture of sucrose and maltose and the resulting complex mixture of oligosaccharides separated on charcoal columns. The first trisaccharides identified were panose, and O-α-D-glucopyranosyl-$(1 \rightarrow 2)$-O-β-D-fructofuranosyl-β-D-fructofuranoside arising from transglucosylation from maltose and trans-fructosidation from sucrose respectively. The new trisaccharide was proved to be O-α-D-glucopyranosyl-$(1 \rightarrow 6)$-O-α-D-glucopyranosyl-$(1 \rightarrow 2)$-β-D-fructofuranoside, Fig. 5.

Here it appeared that it was synthesized by a transglucosylase acting upon maltose as substrate and using sucrose as acceptor molecule. An analogue of this trisaccharide, namely, O-α-D-glucopyranosyl $(1 \rightarrow 4)$-O-α-D-glucopyranosyl $(1 \rightarrow 2)$-β-D-fructofuranoside came from the action of an invertase in honey upon sucrose (White, 1952).

It was considered of some interest to examine the role of the new trisaccharide in dextran synthesis (Bailey *et al.* 1957). The dextran sucrase from our *Betacoccus arabinosaceus* had no action upon the trisaccharide alone, but when incubated in the presence of sucrose both components disappeared and there appeared free fructose together with a series of oligosaccharides. The oligosaccharides were separated into tetra-, penta- and hexaoses, etc., and it was shown, particularly by infra-red investigation, that they contained isomaltose or $(1 \rightarrow 6)$ linkages. It was considered that they used the trisaccharide as a receptor and not as a substrate and that the synthesis of the oligosaccharides proceeded thus: [glucose-$(1 \rightarrow 2)$-fructose] + [glucose-$(1 \rightarrow 6)$-glucose-$(1 \rightarrow 2)$-fructose]$\rightleftharpoons$[glucose-$(1 \rightarrow 6)$-glucose-$(1 \rightarrow 6)$-glucose-$(1 \rightarrow 2)$-fructose] + fructose and so on, to give penta- and higher saccharides.

Since there is no accumulation of oligosaccharides in normal dextran synthesis it would appear that a single chain mechanism is involved. It is interesting to note that one can usually detect a fructose unit at the end of most dextran chains.

We have made some attempts to investigate the mechanism of the biosynthesis of cellulose by studying the synthesis of β-linked glucosaccharides (Barker, Bourne & Stacey, 1953; Barker, Bourne, Hewitt & Stacey, 1955). We have made a study of the oligosaccharides formed from cellobiose by *Aspergillus niger* (strain 152) using growing cells, resting cells and cell-free extracts. The enzyme system of the mould is able to transfer a glucose unit from one cellobiose to another cellobiose molecule, from cellobiose to other β-linked disaccharides, and from cellobiose to glucose. Among the new linkages formed was the β-$(1 \rightarrow 4)$ or true cellobiose linkage. Various β-linked oligosaccharides were produced. Other workers using *Chaetomium globosum* have produced cellotriose and higher β-linked oligosaccharides from cellobiose (Buston & Khan, 1956).

In our systems pentoses, ketohexoses and N-acetyl-D-glucosamine could function as receptors in the transfer reaction.

Two striking claims have recently been made for the enzymic synthesis of cellulose. In the first process an enzyme preparation was flung down by very high-speed centrifugation of sonically disrupted cells of *Acetobacter xylinum*. A buffered suspension of this, on incubation with ^{14}C-glucose-labelled UDPG, gave rise to radioactive cellulose identified by the isotope-dilution method (Glaser, 1957).

Cellodextrins stimulated the formation of cellulose. The author, using

UDPG glucosamine had similarly, previously, synthesized a chitin-like molecule.

In the second investigation Greathouse disrupted *A. xylinum* cells by a variety of abrasive methods, centrifuged off cell debris and freeze-dried the aqueous solution. A 10 % enzyme solution of this was incubated at 30° with 1 % glucose labelled in various positions with ^{14}C and phosphate buffer together with 0·1 % adenosine triphosphate at a pH optimum of 9. It was claimed that high yields of labelled cellulose were produced within 2–4 hr. (Greathouse, 1957).

In many closely similar experiments carried out previously we have been quite unable to produce any cellulose from glucose by a soluble *Aceto-bacter* enzyme.

The use of enzymes for synthesizing di- and oligosaccharides which may be difficult to make by chemical methods offers high rewards for the future. Recently we have described work involving the action of an enzyme from *Aspergillus niger* on a cellobiose-xylose mixture. There was synthesized, by glycosidic exchange, a glucosyl xylose disaccharide which was identified as 3-*O*-β-D-glucopyranosyl-D-xylose (Barker, Bourne, Hewitt & Stacey, 1957).

The enzymes also provide a means of studying the influence of poly-hydroxy compounds as possible receptor groups (Barker, Bourne, Stacey & Ward, 1957). Thus it was tempting to examine the effect of streptomycin with various carbohydrate-enzyme systems. Some interesting retardations of enzyme action were observed, but no evidence was obtained that streptomycin could be incorporated into polysaccharide structures.

Some interesting aspects of enzymic transglycosylation reactions in oligosaccharide formation are contained in the review by Edelman (1956).

REFERENCES

ALBON, N., BELL, D. J., BLANCHARD, P. H., GROSS, D. & RUNDELL, J. T. (1953). *J. Chem. Soc.* 24.

BACON, J. S. D. & BELL, D. J. (1953). *J. Chem. Soc.* 2528.

BAILEY, R. W., BARKER, S. A., BOURNE, E. J. & STACEY, M. (1954). *Nature, Lond.* **174**, 635.

BAILEY, R. W., BARKER, S. A., BOURNE, E. J. & STACEY, M. (1955). *Nature, Lond.* **176**, 1164.

BAILEY, R. W., BARKER, S. A., BOURNE, E. J., STACEY, M. & THEANDER, O. (1957). *Nature, Lond.* **179**, 310.

BARKER, S. A. & BOURNE, E. J. (1952). *J. Chem. Soc.* 209.

BARKER, S. A. & BOURNE, E. J. (1953). *Quart. Rev.* **7**, 56.

BARKER, S. A., BOURNE, E. J. & CARRINGTON, T. R. (1954). *J. Chem. Soc.* 2125.

BARKER, S. A., BOURNE, E. J., GRANT, P. M. & STACEY, M. (1956). *Nature, Lond.* **178**, 1221.

BARKER, S. A., BOURNE, E. J., HEWITT, G. C. & STACEY, M. (1955). *J. Chem. Soc.* 3734.

BARKER, S. A., BOURNE, E. J., HEWITT, G. C. & STACEY, M. (1957). *J. Chem. Soc.* 3541.

BARKER, S. A., BOURNE, E. J. & STACEY, M. (1953). *Chem. Ind.* 1287.

BARKER, S. A., BOURNE, E. J., STACEY, M. & WARD, R. B. (1957). *J. Chem. Soc.* 2944.

BEALING, F. J. & BACON, J. S. D. (1951). *Biochem. J.* **49**, lxxv.

BUSTON, H. W. & KHAN, A. H. (1956). *Biochim. Biophys. Acta,* **19**, 564.

CORI, C. F. (1945). *Fed. Proc.* **4**, 226.

CORI, G. T., SWANSON, M. A. & CORI, C. F. (1945). *Fed. Proc.* **4**, 234.

EDELMAN, J. (1956). *Advanc. Enzymol.* **17**, 189.

GLASER, L. (1957). *Biochim. Biophys. Acta,* **25**, 436.

GREATHOUSE, G. (1957). *J. Amer. Chem. Soc.* **79**, 4503.

HANES, C. S. (1940a). *Proc. Roy. Soc.* B, **128**, 421.

HANES, C. S. (1940b). *Proc. Roy. Soc.* B, **129**, 174.

HASSID, W. Z. & DOUDOROFF, M. (1950). *Advanc. Enzymol.* **10**, 123.

HAWORTH, W. N., PEAT, S. & BOURNE, E. J. (1944). *Nature, Lond.* **154**, 236.

HEHRE, E. J. (1951). *Advanc. Enzymol.* **11**, 237.

KOEPSELL, H. J., TSUCHIYA, H. M., HELLMAN, N. N., KAZENKO, A., HOFFMANN, C. A., SHARPE, E. S. & JACKSON, R. W. (1953). *J. Biol. Chem.* **200**, 793; see also *J. Amer. Chem. Soc.* (1953), **75**, 757.

LELOIR, L. F. & CARDINI, C. E. (1955). *J. Biol. Chem.* **214**, 149.

MONOD, J. & TORRIANI, A. M. (1950). *Ann. Inst. Pasteur,* **78**, 65.

PEAT, S., WHELAN, W. J. & HINSON, K. A. (1955). *Chem. Ind.* 385.

SHIBASAKI, K. & ASO, K. J. (1955). *Ferment. Technol. Japan,* **33**, 45.

STACEY, M. (1954). *Advanc. Enzymol.* **15**, 301.

STACEY, M. (1957). *Biochimia,* **22**, 241.

STODOLA, F. H., KOEPSELL, H. J. & SHARPE, E. S. (1952). *J. Amer. Chem. Soc.* **74**, 3202.

WHELAN, W. J. (1956). *Nature, Lond.* **178**, 1221.

WHITE, J. W. (1952). *Arch. Biochem. Biophys.* **39**, 239.

POSSIBLE MECHANISMS BY WHICH INFORMATION IS CONVEYED TO THE CELL IN ENZYME INDUCTION

By M. R. POLLOCK and J. MANDELSTAM

National Institute for Medical Research, Mill Hill, London, N.W. 7

Replica. A copy, duplicate or reproduction of a work of art (*Shorter Oxford English Dictionary*). A duplicate executed by the artist himself (*Standard Dictionary of the English Language*).

In enzyme induction the inducer, which is either the substrate of the enzyme or a chemically similar substance, specifically stimulates synthesis of the enzyme. The inducer may thus be said to convey information causing

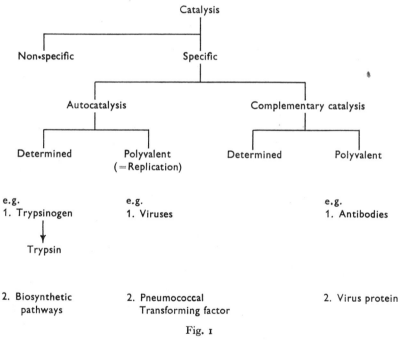

Fig. 1

a specific alteration in the metabolism of the cells. Before considering the nature of this information it will be useful to consider in what general class of biochemical reactions enzyme induction should be included.

In Fig. 1 a tentative general classification of biochemical catalysis is shown and various processes have been assigned their places in the scheme. In this context we are concerned only with specific catalysis, which we have subdivided into autocatalysis and complementary catalysis.

AUTOCATALYSIS

A biochemical process is defined as autocatalytic when the addition of a small amount of a molecular species—the prototype—leads to the formation of more of these molecules. A relatively simple example is the conversion of trypsinogen to trypsin which is initiated by a trace of trypsin and then proceeds to completion (Northrop, 1937). This is classified as a determined reaction because the system has no freedom of copy, i.e. it can only produce trypsin molecules. Similarly a micro-organism may require a trace of a metabolite to initiate growth, after which growth proceeds satisfactorily, accompanied by synthesis of more molecules of the factor concerned. Here again there is a determined biosynthetic pathway which can only produce the prototype molecule. Addition of some other prototype molecule such as an analogue of the growth factor would not lead to synthesis of more of the analogue even if it did serve to initiate growth. Because there is no freedom of copy this is merely a more complex example of determined autocatalysis.

In a polyvalent system, however, there is freedom of copy, i.e. the *same* system has the potential ability to copy more than one species of prototype molecule, and what it actually produces is determined by the prototype which initiates the process. It seems essential to restrict the term 'replication' to polyvalent autocatalysis in order to distinguish between true copying mechanisms and simple 'trigger' reactions like the trypsinogen-trypsin conversion. It must be emphasized that when we say that the 'same' system can copy several species of molecule it means only that, in the present state of knowledge or level of analysis, it cannot be shown that there is a separate determined pathway for each species of molecule the system can produce. For instance, the production of T2 or T4 phage by *Escherichia coli*, according to whether it is infected by T2 or T4, must, at the cellular level, be considered as an instance of polyvalent autocatalysis. But, although it is unlikely, the possibility cannot be excluded that there are separate specific phage-producing systems within the cell, each capable of forming one species of phage when presented with the requisite information. If this were so it would mean that, at a subcellular level, phage production would have to be considered as an example of determined autocatalysis.

COMPLEMENTARY CATALYSIS

In complementary catalysis the presence of the prototype molecule leads to the formation of molecules having a structure which is in some way complementary to that of the prototype. The complementary nature of the process is apparent when the prototype and the product can react chemically as is the case with antigen and antibody, or substrate and enzyme. Similarly the template theories of protein synthesis assume that a complementary relationship exists between ribonucleic acid and protein.

Again it is useful to apply the criterion of freedom of copy to distinguish between determined and polyvalent reactions. The phenomenon which probably corresponds most closely to the polyvalent type of complementary catalysis is antibody formation. Here a large number of substances may serve as prototypes. There is, however, some restriction on freedom of copy, since not all foreign macromolecules act as antigens, and there are considerable differences in antibody response between different species of animals and also between different individuals of the same species. In spite of this there does not appear to be any limit to the number of substances that can act as prototypes.

Examples of other polyvalent reactions are the ability of virus nucleic acid to cause production of the specific protein molecules which are characteristic of that virus. It may be noted in passing that, when virus synthesis is initiated by infective nucleic acid, autocatalysis and complementary catalysis proceed together. The nucleic acid causes the replication of itself and the complementary production of specific proteins.

If complementary catalysis occurs reciprocally between two species of molecule, the result is an autocatalytic process. Schematically,

$$\text{metabolites} \xrightarrow{A} B$$

$$\text{metabolites} \xrightarrow{B} A$$

i.e. A catalyses complementary synthesis of B and vice versa. Addition of either A or B will result in the production of more of both. It may well be that some of the reactions we have classified as replication, may be found to involve reciprocal complementary catalysis.

ENZYME INDUCTION

In the process of enzyme induction the prototype molecule—either the substrate or a related compound—specifically evokes the formation of a protein molecule having a structure which is, in part at any rate, complementary to that of the prototype. This complementary character of the reaction, and the fact that a number of different kinds of enzyme may be inducible in the same organism, would seem at first sight to place enzyme induction in the category of polyvalent reactions, and it has indeed often been compared with antibody formation.

It is the purpose of this paper to discuss whether it is justifiable to class enzyme induction as a polyvalent process or whether it should fall in the category of determined biosynthetic pathways. In other words we have to consider whether the inducer carries information determining the properties of a protein molecule in the same sort of way that an antigen molecule is commonly supposed to do—namely as a pattern or template upon which the structure of the induced molecule is moulded. (For the purpose of this discussion it does not matter whether the information concerns the amino acid sequence of the protein or merely the manner in which a pre-existing sequence is folded.)

In spite of the superficial similarity between enzyme induction and antibody formation some differences are immediately apparent. First, antibody formation, as we have already noted, can be evoked by an apparently unlimited number of substances while in enzyme induction the possibilities are more restricted and genetically determined (see below). Secondly, the prototype molecule in the antibody reaction is usually a macromolecule, while in enzyme induction it is usually a small molecule. This is, however, not an essential difference since some inducible enzymes (e.g. amylase) may be formed in response to macromolecular inducers. A more important difference lies in the fact that the antibody reaction apparently leads to the appearance of a new species of molecule while induction does not. (This point will be considered in detail below.) It is, of course, impossible to be certain that an animal does not already possess a few molecules of antibody to every substance that could act as an antigen. However, in at least one case where natural ('pre-immunization') antibodies are detectable, they have been shown to differ physico-chemically from the induced antibodies developed after immunization (Mackay, 1957). On the present evidence, it seems reasonable to suppose that the structure of the antigen, directly or indirectly, determines the structure of the antibody and leads to the appearance of a new molecular pattern in the organism.

In the following sections, evidence will be adduced to show that the process of enzyme induction is so greatly restricted that it does not fall into the category of polyvalent reactions, and that the kind of information conveyed by the inducer differs fundamentally from the kind of information conveyed by the antigen.

GENETIC RESTRICTIONS ON INDUCIBILITY

The potentialities of micro-organisms to form inducible enzymes are rigidly restricted and can be shown to be genetically determined (Lederberg, 1948). In yeast, the potential capacity to form induced enzymes is inherited in a strictly Mendelian fashion (see e.g. Lindegren, Spiegelman & Lindegren, 1944). Similarly the ability of *Bacillus cereus* to form penicillinase by induction or to form it constitutively (i.e. without the necessity of adding penicillin) is also genetically determined (Sneath, 1955). Again, it has long been known that certain strains of *Escherichia coli* which are unable to utilize lactose at all, spontaneously produce a small proportion of mutants which can utilize it by forming an inducible enzyme (Lewis, 1934).

In many of these different systems the transition from one type to another has been shown to be the result of a single mutational step. In other words, whether the cells will react to addition of an inducer, or indeed whether they require an exogenous inducer at all, is dependent upon genetic 'information' already contained in the system.

THE BASAL ENZYME

Where sufficiently refined methods of enzyme assay have been used, it has usually proved possible to show that micro-organisms already contain a small amount of enzyme before treatment with the inducer. This low level of enzyme produced in a constitutive manner has been called the basal enzyme. A study of biochemical properties and serological reactions alike indicates that the enzyme produced in response to exogenous inducer is identical with that produced spontaneously (Monod & Cohn, 1952; Kogut, Pollock & Tridgell, 1956).

The synthesis of the basal enzyme has not been investigated as intensively as synthesis of induced enzyme and, at this stage, one can only speculate upon the manner of its production. One of the simplest assumptions to make is that the mechanism is the same as that involved in induced synthesis, i.e. the basal level of the enzyme is the result of the presence of an endogenous inducer. This would be a likely explanation for the high basal level of lysine decarboxylase in *Bacterium cadaveris* (Mandelstam, 1954) since lysine is a normal metabolite of the cells and its presence in the free state can be demonstrated.

In many other instances, the inducer is not known to be a metabolite (e.g. penicillin) nor has its presence in the cells been demonstrated. In practice it is difficult to disprove the existence of an endogenous inducer since only one or two molecules per cell, acting catalytically, might well be sufficient to account for the basal level of enzyme. The alternative, and more likely explanation, is that the basal enzyme is produced without the intervention of endogenous inducer.

If we adopt this explanation it is clear that the cells already possess the 'information' required to produce the specific protein pattern of the inducible enzyme. The addition of exogenous inducer does not in other words result in the formation of a new species of molecule; instead its effect is to accelerate the production of a protein which the cells can produce in any case.

MODE OF ACTION OF INDUCERS

In earlier attempts to explain induction it was assumed that the enzyme was produced continuously but was unstable, and the function of the inducer was merely to combine with the enzyme and so protect it from inactivation (Yudkin, 1938; Spiegelman, 1946; Monod, 1947). More recent work has shown that the inducer can react with some substance other than the enzyme and that its function is probably catalytic.

The possible mode of action of the inducer has been discussed at length in a number of reviews (Monod & Cohn, 1952; Mandelstam, 1956; Pollock, 1958) and the facts can only be briefly summarized here.

Evidence from the β-galactosidase and penicillinase systems suggests that the inducer (I), or some specific metabolite derived from it, enters into combination with a specific receptor molecule (R) to form an organizer (O). The scheme may be represented as follows:

$$I + R$$
$$\downarrow$$
$$O$$
$$\text{amino acids} \longrightarrow \text{enzyme}$$

The organizer then catalyses the synthesis of the enzyme from amino acids. The catalytic nature of the process can be demonstrated in the case of penicillinase. In this system the inducer is specifically and irreversibly fixed after transient contact with cells of *Bacillus cereus*. All free penicillin can then be removed, and the bacteria allowed to synthesize enzyme. It can be shown that about 40 molecules of enzyme are produced per hour per molecule of inducer originally fixed by the cells. The specificity of the penicillin receptor is confirmed by the fact that there is competitive interaction for attachment between penicillin and its analogues cephalosporin C and cephalosporin N. It thus appears that the uninduced cells possess a limited number of receptor molecules (about 1000 per cell) which have a specificity very similar to that of penicillinase itself.

CLASSIFICATION OF ENZYME INDUCTION

Summing-up the facts so far discussed we find that the ability to synthesize a particular inducible enzyme is genetically determined. In addition, the existence of the basal enzyme and the specific receptor indicate that no new pattern is introduced into the cell by adding inducer. The inducer merely increases the rate of production of protein molecules by an existing biosynthetic pathway.

It therefore seems reasonable that enzyme induction should be placed in the class of determined biosynthetic pathways rather than in the class of polyvalent or copying reactions. From this point of view the function of the inducer would be to act as part of a rate-controlling mechanism and not as part of a template.

THE CONTROL OF METABOLITE CONCENTRATION

From recent work it would appear that the specific effect of small molecules upon the rate of enzyme synthesis is not limited to enzyme induction but may be one of the means whereby the concentration of intermediary metabolites is controlled.

Let us consider the means whereby the concentration of a metabolite in a metabolic sequence might be automatically controlled.

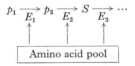

The metabolite S is formed from precursors p_1 and p_2 by the action of enzymes E_1 and E_2; it is destroyed by E_3. It is clear that S could control its own concentration if it could decrease its rate of formation or increase its rate of destruction, e.g. it might specifically inhibit production of E_1 or E_2, or it might specifically stimulate production of E_3 (i.e. enzyme induction). (In theory, the same result could be achieved by affecting the action of the enzymes rather than their production, and examples of this have been described. This method of controlling metabolite concentration is outside the scope of this paper.)

Enzyme induction is, therefore, only one particular instance of the way in which a small molecule could specifically affect the rate of formation of an enzyme. Examples of specific inhibition have indeed been described but have not been so extensively studied as induction. It is interesting to note that all the types of control that are theoretically possible can be found in practice.

(1) The metabolite inhibits the formation of the enzyme immediately responsible for its production (i.e. inhibition of E_2 synthesis).

An example is the tryptophan synthetase of *Aerobacter aerogenes* which catalyses the reaction

$$\text{indole} + \text{serine} \rightarrow \text{tryptophan}$$

Monod & Cohen-Bazire (1953) showed that synthesis of the enzyme was fairly specifically inhibited by the addition of tryptophan to the growth medium. Similarly the production of methionine-synthetase in *Escherichia coli* was inhibited in the presence of methionine (Cohn, Cohen & Monod, 1953).

(2) The metabolite inhibits production of an enzyme further back in the sequence (inhibition of E_1 synthesis).

An example of this is to be found in the metabolic pathway from glutamic acid to arginine in *E. coli*. The later stages of the sequence are as follows (Abelson, 1954):

$$N^\alpha\text{-acetylornithine} \rightarrow \text{ornithine} \rightarrow \text{citrulline} \rightarrow \text{arginine}$$

It has been shown (Vogel, 1957) that the addition of arginine to the growth medium inhibits the synthesis of N^α-acetylornithinase.

Suppression of synthesis in this kind of way may be a fairly common means of metabolic control. Thus, Roberts *et al.* (1955), working with *E. coli*, examined the effect of fourteen amino acids upon the rate of their own production. In the following nine cases endogenous synthesis from glucose and ammonium salts was almost totally suppressed: proline, serine, methionine, threonine, arginine, cystine, lysine, leucine and isoleucine. The other five amino acids, valine, alanine, glycine, glutamic acid and aspartic acid did not suppress their own synthesis.

This type of control is not confined to amino acids. The synthesis of purines in *E. coli* is inhibited by addition of purines to the growth medium (Koch, Putnam & Evans, 1952).

In these instances the mechanism of inhibition is unknown and it is possible that some may turn out to be cases of suppression of enzyme action instead of enzyme production. For instance, this has been shown to be the case with isoleucine, which is formed by a series of reactions from threonine (Abelson, 1954), and which inhibits the action of threonine deaminase several steps back in the metabolic sequence (Umbarger, 1956).

The metabolism of arginine in *E. coli* provides an instance where two mechanisms may operate simultaneously to control the concentration of the metabolite, for not only can arginine inhibit its own synthesis, but in many strains it will induce arginine decarboxylase under suitable conditions

(Gale, 1940), and so increase its own rate of removal. Thus, if *E. coli* is grown in glucose and ammonium salts we have

$$\text{glucose} + NH_4 + \cdots\cdots\rightarrow N^\alpha\text{-acetylornithine} \cdots\cdots\rightarrow \text{arginine}$$

$$\downarrow \text{ to protein}$$

If arginine is added to the medium,

$$\text{agmatine}$$
$$\uparrow \text{ arginine}$$
$$| \text{ decarboxylase}$$
$$\text{glucose} + NH_4 + \cdots\cdots\rightarrow N^\alpha\text{-acetylornithine} \cdots | \cdots\rightarrow \text{arginine}$$

$$\downarrow \text{ to protein}$$

The synthesis of arginine is retarded and the induced synthesis of arginine decarboxylase is stimulated.

WHERE DOES THE INDUCER ACT?

If it is accepted that the effect of the inducer is an effect upon the rate of production, we have to enquire how this effect might be exerted, i.e. whereabouts in the process of protein synthesis could the inducer intervene?

Here, where the end-product of the reaction is a protein, the problem is obviously more complicated than in the case of a simple metabolite. Nothing like a reaction sequence has been established.

So far as induced enzymes are concerned the evidence from isotopic experiments indicates that they, like other proteins, are formed directly from amino acids (Rotman & Spiegelman, 1954; Hogness, Cohn & Monod, 1955; Pollock, 1957). So at the moment we can only formulate the reaction very generally as follows:

$$\text{Precursors (?nucleotides)}$$
$$\downarrow$$
$$\text{Amino acids} \xrightarrow{\text{enzyme-forming site}} \text{enzyme}$$
$$\text{(E.F.S.)}$$

If the enzyme acted like some of the simpler metabolites we have considered it might itself control its rate of synthesis. This could be achieved by inhibiting either the action or the formation of the enzyme-forming site (E.F.S.). On the other hand it is possible that E.F.S. is inhibited in some other way. In either case, the function of the inducer would be to relieve the inhibition. This could be achieved by combining with the endogenous inhibitor or alternatively by competitively occupying the site at which the endogenous inhibitor acts. The result would be the freeing of a catalytic surface from inhibition so that the inducer might be termed a 'pseudo-catalyst'.

This interpretation is purely speculative and it will perhaps not be possible to define the function of the inducer in greater detail until more is known of the manner in which amino acids are assembled at the sites of protein synthesis.

REFERENCES

ABELSON, P. H. (1954). *J. Biol. Chem.* **206**, 335.

COHN, M., COHEN, G. N. & MONOD, J. (1953). *C.R. Acad. Sci., Paris*, **236**, 746.

GALE, E. F. (1940). *Biochem. J.* **34**, 392.

HOGNESS, D. S., COHN, M. & MONOD, J. (1955). *Biochim. Biophys. Acta*, **16**, 99.

KOCH, A. L., PUTNAM, F. W. & EVANS, E. A. (1952). *J. Biol. Chem.* **197**, 105.

KOGUT, M., POLLOCK, M. R. & TRIDGELL, E. J. (1956). *Biochem. J.* **62**, 391.

LEDERBERG, J. (1948). *Genetics*, **33**, 617.

LEWIS, I. M. (1934). *J. Bact.* **28**, 619.

LINDEGREN, C. C., SPIEGELMAN, S. & LINDEGREN, G. (1944). *Proc. Nat. Acad. Sci., Wash.* **30**, 346.

MACKAY, M. E. (1957). *Biochem. J.* **66**, 545.

MANDELSTAM, J. (1954). *J. Gen. Microbiol.* **11**, 426.

MANDELSTAM, J. (1956). *Int. Rev. Cytol.* **5**, 51.

MONOD, J. (1947). *Growth*, **11**, 223.

MONOD, J. & COHEN-BAZIRE, G. (1953). *C.R. Acad. Sci., Paris*, **236**, 530.

MONOD, J. & COHN, M. (1952). *Advanc. Enzymol.* **13**, 67.

NORTHROP, J. H. (1937). *Physiol. Rev.* **17**, 144.

POLLOCK, M. R. (1957). *International Symposium on Enzyme Chemistry* (in the Press).

POLLOCK, M. R. (1958). In *The Enzymes* (in the Press). New York: Academic Press.

ROBERTS, R. B., ABELSON, P. H., COWIE, D. B., BOLTON, E. T. & BRITTEN, R. J. (1955). *Studies of Biosynthesis in* Escherichia coli. Washington, D.C.: Carnegie Institution Publication 607.

ROTMAN, B. & SPIEGELMAN, S. (1954). *J. Bact.* **68**, 419.

SNEATH, P. H. A. (1955). *J. Gen. Microbiol.* **13**, 561.

SPIEGELMAN, S. (1946). *Cold Spr. Harb. Symp. Quant. Biol.* **11**, 256.

UMBARGER, H. E. (1956). *Science*, **123**, 848.

VOGEL, H. J. (1957). In *The Chemical Basis of Heredity*, p. 276. Ed. McElroy, W. D. & Glass, B. Baltimore: Johns Hopkins Press.

YUDKIN, J. (1938). *Biol. Rev.* **13**, 93.

PROCESSES CO-ORDINATING INTRACELLULAR ACTIVITY

By ALFRED MARSHAK*

Marine Biological Laboratory, Woods Hole, Mass.

Replication of macromolecules takes place within cells and for this reason it seems appropriate that it be considered in the perspective of our understanding of the regulatory mechanisms of the cell. Some, which we have been aware of for a long time, will only be briefly mentioned, although this by no means implies that they are fully understood. About others we have limited and conflicting information, so that decisions as to their function are matters of choice. I shall introduce a third category that is almost entirely hypothetical, but which seems to be needed to account for existing observations.

The differential accumulation of ions and molecules by the cell through the intervention of a semi-permeable membrane, together with processes requiring the expenditure of energy, may have its intracellular counterpart in the behaviour of mitochondria (Höber, 1945; Krogh, 1946; Watanabe & Williams, 1953; Lindberg & Ernster, 1954). The nucleus has been found to show osmotic behaviour and in addition to display an electro-osmotic phenomenon in which both nuclear envelope, and a cytoplasmic constituent joined to it, act together to regulate the inward flow of some substances (Marshak, 1957). Apart from membrane separations, the distribution of substances in the cytoplasm depends not only on diffusion rates and chemical affinities, but also on a mixing process, cyclosis, and on molecular sorting phenomena like 'intracellular cataphoresis' (Spek, 1930, 1934), and the concentration of some dyes into minute vacuoles (Spek & Chambers, 1933). The existence in the cytoplasm of living cells of minute 'vacuoles' was evident from studies made quite a number of years ago with the light microscope (Baas-Becking, Bakhuysen & Hotelling, 1928; Seifriz, 1942). The electron microscope has, of course, revealed a wealth of detail in submicroscopic cellular structures, some of the functions of which are beginning to be known. The enzymes involved in the transfer of electrons from substrate to oxygen and others, less well known, concerned with phosphorylations, are contained in the mitochondria (Lehninger, 1956). Much of the cytoplasmic ribonucleic acid is contained in the microsomes which seem to be engaged in protein synthesis (Hoagland & Zamecnik, 1957; Zamecnik &

* Aided by a grant from the American Cancer Society.

Keller, 1954). We lack the information needed to give meaning to this distribution of functions. Our ignorance looms large again in the matter of permeability and molecular transport. By what means do large nucleo-protein molecules, or aggregates of them, penetrate into cells and become incorporated in the chromosome apparatus (Jacob & Wollman, 1957; Marshak & Walker, 1945)? Special mechanisms may be invoked for the injection of phage nucleic acid but not for the penetration of particles of chromatin into liver cells, where entrance into the cell may be by pinocytosis or other unknown means. Once in the cell, bacterial or mammalian, approach to the nucleus and incorporation into the chromosomes would scarcely seem to be a matter of chance, but for a mechanism we can, at present, only suggest that it may involve movements like those shown by the chromosomes themselves.

We know that the capacity of the cell for carrying out metabolic trans-formations is dependent upon its complement of constitutive enzymes and its capacity for generating a variety of adaptive enzymes. Chemical trans-formations by enzymatic reactions are known to occur by alternative pathways. A particularly interesting one is that in which the branching involves a feed-back mechanism such as that postulated by Gots & Gollub for nucleotide synthesis (1957). From experiment as well as theory, there is ample evidence that control of the production of enzymes is exercised by the genetic apparatus of the cell (Horowitz & Fling, 1956). But though the search has been diligent, the products and the reactions which intervene between the gene and the final product, the enzyme, can only be suggested. A possible step in this direction may be found in the analysis of the be-haviour of blebs, in regions of the nuclear membrane in *Drosophila* salivary glands to which a particular portion of a chromosome is attached. Gay has postulated that the blebs, when released into the cytoplasm, become ergastoplasm and that the phenomenon is one of transport of chromosomal products into the cytoplasm (Gay, 1956). While we may lack clues to the nature of the initial gene product, we are acquiring evidence on the manner of intracellular protein synthesis. Some time ago, the old correlation between increased basophilia and synthetic activity was made more specific by the identification of the basophil substance with ribonucleic acid (RNA). Work with bacterial residues has provided further indications of a role played by one or both categories of nucleic acids in protein synthesis (Gale & Folkes, 1955a, b; Spiegelman, 1957). Other experiments with mam-malian microsomes have demonstrated that nucleotide polyphosphates take part in the activation of amino acids which are incorporated into protein (Hoagland, 1955; Hoagland & Zamecnik, 1957). There have also been suggestions of intranuclear protein synthesis in association with deoxyribonucleic acid (DNA) (Allfrey, Mirsky & Osawa, 1957; Irvin, Rotherham, Irvin & Holbrook, 1957). The experiments of Allfrey *et al.* with isolated nuclei, like those of Gale & Folkes and of Spiegelman with bacterial residues, were conducted on systems in a state of progressive deterioration, so that it would be difficult to say whether increased incorporation of a labelled amino acid into a protein fraction, following the addition of a compound such as DNA, was due to stimulation of protein synthesis by the added compound, or whether the compound, by inhibiting degradative reactions, enabled other constituents of the system to carry out more protein synthesis than was possible in its absence. In this connexion it seems significant that such apparent synthesis was stimulated by partially digested DNA and by RNA, as well as by intact DNA. In these *in vitro* experiments with alanine-^{14}C as the administered amino acid, histone was only very poorly labelled (Allfrey *et al.* 1957). In the experiments of Irvin *et al.*, where glycine-^{14}C was administered *in vivo*, the histone became strongly labelled. Comparison was made of the specific activities observed in RNA, DNA, and several protein fractions isolated from liver and hepatoma. The ratio of tumour to liver specific activity was 1 or less in all except the DNA and histone fractions, where it was greater. The authors have suggested that their observations support Kacser's hypothesis for gene and chromosome structure which postulates simultaneous synthesis of DNA and histone (Kacser, 1956). However, the same data would support other models for the gene.

The pathways for the *de novo* synthesis of the purines and pyrimidines have been worked out in considerable detail by Buchanan, Greenberg and their co-workers (Buchanan, 1952; Greenberg & Spillman, 1956; Goldthwait, Greenberg & Peabody, 1955; Levenberg & Buchanan, 1956; Remy, Remy & Buchanan, 1955; Hartman, Levenberg & Buchanan, 1956; Williams & Buchanan, 1953). It is of considerable interest that the purines and pyrimidines are not synthesized as the free bases but as the ribose phosphate derivatives. Analysis of the acid-soluble fraction of various types of organisms and tissues has revealed the presence of the mono-, di-, and tri-5′-phosphates of all four ribonucleosides as well as some more complex derivatives (Cohn, 1956; Hecht, Brumm & Potter, 1957; Hurlbert, Schmitz, Brumm & Potter, 1954; Keller, Zamecnik & Loftfield, 1954; Park, 1952; Schmitz, Hurlbert & Potter, 1954; Zamecnik & Keller, 1954). The di- and triphosphates of the pyrimidine deoxyribosides have been isolated from thymus extracts, and conditions for their *in vitro* formation have been described (Hecht, Potter & Herbert, 1954; Potter & Schlesinger, 1955; Sable, Wilber, Cohen & Kane, 1954). A bacterial enzyme has been isolated which catalyses the polymerization of ribonucleotides from the 5'-nucleoside diphosphates, and some of the polymers so produced have a composition resembling that of RNA (Grünberg-Manago & Ochoa, 1955; Grünberg-Manago, Oritz & Ochoa, 1955, 1956; Ochoa & Heppel, 1957). Another enzyme has been found which leads to the polymerization of deoxyribonucleotides from the deoxyribonucleoside—5'-triphosphates in the presence of DNA (Kornberg, 1957). It has been suggested that intracellular synthesis of RNA and DNA may be brought about by means of enzymes such as these (Kornberg, 1957; Ochoa & Heppel, 1957). In such systems the RNA or DNA may be considered as merely reservoirs of stored energy. However, if the genes are presumed to be composed of DNA, RNA or both, we find in these observations the elements needed for a self-reproducing system which, superficially at least, appears not unlike that which has been suggested by the data of biochemical genetics. Thus the genome initiates the production of enzymes which are concerned with the *de novo* synthesis of the sub-units of the genes, and these are assembled by other enzymes similarly produced into the polymerized products which are the presumed genes themselves, or aggregates of them. The element of control of self-reproduction seems remote, if not missing, in such a scheme. Indeed as far as RNA is concerned, the final genic product would be determined by the composition of the nucleotide pool and by enzyme activity rather than by the parent gene. In Kornberg's system, undenatured DNA is required, so that a possible template action may exist, but only with the intervention of one or more enzymes.

The concept of a template as applied to gene reproduction has long been in existence, but a precisely defined model was not produced before that proposed by Watson & Crick, which in addition to satisfactorily accounting for X-ray diffraction data and for the proportions of bases in DNA, provided a scheme whereby polynucleotide chains, associated in pairs through hydrogen-bond linkage of the 6-amino and 6-keto groups of the bases, could accurately reproduce themselves. Extrapolating beyond the chemical and diffraction data, it has been proposed that DNA and DNA alone is the genic substance (Crick & Watson, 1954; Watson & Crick, 1953). Difficulty with the model has been encountered in the matter of attempting to fit into it amounts of 6-methylaminopurine such as have been found in some *Escherichia coli* mutants (Crick, 1957; Dunn & Smith, 1955). Bendich, Pahl & Brown (1957) reported another chemical observation which seemed inconsistent with the model in that fractions of DNA obtained by chromatographic separation showed base ratios which differed significantly from those required by the model. Although the model provides neatly for its self-duplication, there is a possible snag in the problem of the separation of the pairs of plectonemically coiled chains, which is left unsolved. Investigations by Plaut & Mazia (Mazia, 1956) with chromosomes labelled with thymidine-^{14}C seemed to give results at variance with the predictions of the Watson & Crick model, but subsequent experiments with thymidine-^{3}H by Taylor, Woods & Hughes (1957) in which the radioactivity could be followed much more precisely, gave results in agreement with the hypothesis. The model makes no predictions about the nature of the agents intervening between the presumed DNA genes and their cytoplasmic products. The results of Taylor *et al.* have a resemblance to those of Levinthal (1956), in which a section of the phage T2 hereditary material equivalent to about 40 % of the parental DNA was transmitted to the first-generation progeny, and half of that to the second generation. However, the fact that 40 % of the DNA-^{32}P was transmitted as a unit does not mean that the DNA in this section remained intact. It may, for example, have been held together by being bound to a 'substratum' of protein or RNA, or both. In any case, the intervention of RNA or protein as possible precursors of new phage or new chromosome fibrils is not excluded by these experiments.

If we are to equate genes and DNA, it follows that the DNA must retain its structure, i.e. it must remain in the polymerized form throughout the life-history of any cell that retains its capacity for reproduction. This it apparently fails to do in some species of echinoderm eggs. It has been found that although the chromosomes of these eggs remain strongly Feulgen-positive through both polar body divisions, and though the polar body nuclei also remain Feulgen-positive, the egg pronucleus becomes Feulgen-negative. The negative reaction could not be accounted for by dispersion of the DNA in a large volume, since the pronucleus was much smaller than the strongly Feulgen-positive nuclei of the early meiotic prophase stages; nor could it be explained away by postulating unusual diffuse or branched chromosomes, for when stained with haematoxylin or examined by phase contrast, the chromosomes had the usual appearance (Marshak & Fager, 1950; Marshak & Marshak, 1953). But a staining reaction, even a specific one, cannot give a definitive answer if the reaction is negative. Therefore quantitative determinations of the thymine content of preparations of eggs of two species of echinoderms were made using the isotope dilution method with uniformly labelled thymine-^{14}C. In spite of all precautions, somatic cells as well as polar bodies were found in the egg suspensions. In one species, *Arbacia*, the somatic contaminants were clumped into spherical masses, and the amount present could be determined only approximately; but in the other species, *Tripneustes*, they were dispersed, so that their number could be determined accurately. In the

Arbacia experiments, the eggs were not extracted with cold dilute acid prior to KOH digestion and acid precipitation of the DNA and protein, but the *Tripneustes* eggs were so extracted. In *Arbacia* all of the thymine found could be accounted for as a constituent of the non-egg cells, except for an amount equal to about five times the amount per sperm cell. When another batch of these eggs was extracted with cold dilute acid, the extract was found to contain thymine in an amount equivalent to 4·85 times the amount per sperm cell. In other words, almost all of the thymine found could be accounted for as DNA of the polar bodies and somatic cells, and as acid-soluble thymine. In the case of *Tripneustes*, the thymine found after acid extraction could be accounted for entirely by the polar bodies and the somatic contaminants, leaving none that could be assigned to the eggs themselves. The *maximum* error for all of the operations in the *Tripneustes* experiments was equivalent to 6 % of the thymine per sperm. Thus the limits of error of the experiments allowed the assignment to the egg of an amount of DNA equivalent to not more than a few per cent of the haploid quantity, if any at all. The loss of Feulgen stainability, the disappearance of all, or almost all, of the DNA, the appearance of acid-soluble thymine, and the failure to obtain Feulgen-positive aggregates in centrifuged echinoderm pronuclei; were all observations considered to be not merely coincidental, but different aspects of the same phenomenon, namely the degradation of the DNA of the eggs at maturation to units of small molecular size (Marshak & Marshak, 1953; 1955a, b; Marshak, 1954).

In another sea urchin, *Strongylocentrotus droebachiensis*, and in the star-fish, *Asterias forbesii*, the same nuclear Feulgen negativity on maturation has been observed (Marshak, unpublished). In the plant *Aloe davyana*, it has been found that the egg nucleus is Feulgen-negative although the other nuclei associated with it in the embryo sac are Feulgen-positive (Krupko & Denley, 1956). In *Habrobracon* eggs irradiated with X-rays, the nuclei of the first few embryo divisions are Feulgen-positive, but nuclei produced in subsequent divisions are Feulgen-negative, both types being found together in the same ooplasm (von Borstel, 1955). The pronucleus of the mouse ovum is Feulgen-positive, as are the polar body nuclei, and it remains so until just before fusion with the male pronucleus, when it becomes Feulgen-negative. There is no appreciable enlargement of the pronucleus and significantly a stage intervenes in which Schiff-positive material appears in the nucleus, as would be expected if DNA were being degraded. During this time the polar body nuclei remain Feulgen-positive (Ludwig, 1954). These cytological observations do not in themselves prove that the DNA of the egg is completely degraded. They do, however, support this hypothesis, and show that the phenomenon of DNA degradation before nuclear fusion appears to be widespread. Its appearance at the beginning of the life cycle just before the fusion of the gamete nuclei is probably significant. The observations are not consistent with the hypothesis that the DNA is the sole genic material, for it is evident that whatever role the DNA may subsequently play in the egg nucleus, DNA must be introduced into the chromosomes by other chromosomal constituents when parthenogenetic development is induced in echinoderm and mouse eggs. It may be argued that such experiments do not rule out the possibility of as much as 6 % of the DNA remaining undegraded, and that this is the genetic material. Such a contention would leave unanswered the question of the significance of the other 94 % of the DNA, and would require the subsidiary postulate that there are at least two kinds of DNA, one with a genic function, and the other without.

Ten years ago it was generally accepted that there were only two nucleic acids, RNA and DNA. At that time I reported that the rate of turnover of the RNA of the nucleus (n-RNA) was about fifteen times faster than that of the cytoplasm (c-RNA), and concluded that the two were functionally different. Later it was shown that the n-RNA and the c-RNA differed in their purine and pyrimidine content. From the results of tracer studies, it could be shown that the n-RNA could act as the precursor of the c-RNA but that the converse was not possible (Marshak, 1948; Jeener & Szafarz, 1950). Evidence supporting this hypothesis was obtained from the observation that the time-course for the RNA-specific activities of the cytoplasmic particles and the nuclei was that predicted; also in the finding that each of the constituent nucleotides of the n-RNA had the same ^{32}P content, and that the relative specific activities of the nucleotides remained unaltered with time (Marshak & Calvet, 1949; Marshak & Vogel, 1950). Results obtained with adenine-^{14}C as a tracer supported the conclusions reached with ^{32}P (Fresco & Marshak, 1954). Similar observations were made by Tyner et al. with ^{32}P and glycine-^{14}C and by Smellie et al. (Tyner, Heidelberger & LePage, 1953; Smellie & Davidson, 1956; Smellie, McIndoe & Davidson 1953; Smellie, McIndoe, Logan, Davidson & Dawson, 1953). An analysis of evidence in this connexion is given by Tyner et al., and a review by Smellie (1955). Rose & Schweigert (1953) found evidence that in the synthesis of DNA in the rat, there is a mechanism for the conversion of ribose to deoxyribose in glycosidic linkage. The observations were confirmed and extended by Roll and his co-workers (1956 a, b). It has been suggested that the high turnover rate in the n-RNA and the low rate in c-RNA are coincidental, synthesis in each being by independent pathways (Barnum & Huseby, 1950; Barnum, Huseby & Vermund, 1952, 1953; Hurlbert & Potter, 1952), and the detailed studies of Tyner et al.

did not provide critical evidence excluding either this or the former hypothesis. Brachet & Chantrenne (1956) found that enucleated *Acetabularia* fragments incorporated orotic acid-^{14}C into RNA at a significant though lower rate than the nucleated fragments for quite a time after the enucleation, and cited similar instances of RNA synthesis in non-nucleated cells of other species. They concluded that not all of the *c*-RNA is derived from the *n*-RNA. However, their experiments did not rule out the possibility that RNA synthesis continuing after enucleation might have been due to products left in the cytoplasm by earlier nuclear activity. In amoeba, it has been found that when a ^{32}P-labelled nucleus is introduced into a cell which already contains an unlabelled one, the radioactivity moves in time into the cytoplasm but not into the unlabelled nucleus (Goldstein & Plaut, 1955). This observation supports that portion of the original hypothesis which stated that the *c*-RNA did not function as a precursor of *n*-RNA. From the observation that amoebae lacking a nucleus were unable to maintain their RNA content, Brachet & Chantrenne (1956) came to the conclusion that in this species the nucleus is necessary for the maintenance of the *c*-RNA. Prescott has shown that enucleated amoebae will take up uracil, but will not incorporate it into RNA as will nucleated specimens (Prescott, 1957). Several investigators have argued that since the base composition of *n*-RNA is different from that of *c*-RNA in several tissues and cell fractions, the latter could not be the immediate product of the former (Allfrey *et al.* 1957; Elson, Trent & Chargaff, 1955; Moldave & Heidelberger, 1954). However, this argument seems to be based on a misunderstanding of the hypothesis, for it was not postulated that the change from *n*-RNA to *c*-RNA involved merely a change in the sugar moiety, and in fact evidence was presented for changes in the proportions of the bases (Marshak, 1951*a*). In studies on the incorporation of adenine-^{14}C into RNA, the specific activity/time curves for the total acid-soluble adenine and that of the *n*-RNA showed the former to be still rising while the latter was falling, which is inconsistent with the assumption that all the acid-soluble adenine could serve as precursor to that in *n*-RNA. It was, however, possible that some components of the acid-soluble adenine fraction had a much higher turnover rate than the average, and that these comprised the pool from which the adenine of *n*-RNA was derived (Fresco & Marshak, 1954). Incorporation of labelled ribonucleotides and orotic acid-^{14}C into ribonucleic acid has been described for cell-free and for nuclei-free homogenates (Goldwasser, 1955; Hecht *et al.* 1954, 1957; Hecht & Potter, 1956; Potter, Hecht & Herbert, 1956). However, it was shown that the label incorporated into RNA in this way remained there only for a short period, in contrast to RNA labelled *in vivo*, which, when put into a similar *in vitro* system, retained its label. Furthermore, there was no net RNA synthesis in the homogenate (Potter, Schneider & Hecht, 1957). When labelled orotic acid was injected into rats with regenerating livers, the results were essentially the same as those previously reported for ^{32}P and adenine-^{14}C, incorporation and turnover in n-RNA being much more rapid than in c-RNA (Hecht & Potter, 1956; Yasuyuki, Hecht & Potter, 1956). A considerable body of data thus supports the hypothesis that n-RNA gives rise to c-RNA. Moreover, some experiments, cited as giving results inconsistent with the hypothesis, have alternative interpretations which either have no bearing on the hypothesis or are consistent with it. Moldave & Heidelberger (1954) reported experiments in which *in vivo*-labelled cell particulates were digested with enzymes and with alkali to hydrolyse RNA, the results being taken to prove the intramolecular heterogeneity of the RNA of each of the particle fractions. However, an alternative interpretation can be given in terms of contaminants derived from impurities in the enzymes used, or from inadequate separation of the final products measured. Furthermore the homogeneity of their particulate fractions was not established. While there may very well be 'intramolecular heterogeneity' in RNA biosynthesis, the experiments cited do not demonstrate this condition. Sacks & Samarth (1956) subjected RNA of cytoplasmic particles from ^{32}P-labelled rodent livers to fractional degradation by trichloracetic acid and found non-uniformity in the specific activity of the fractions so obtained. Since the individual nucleotides were not isolated nor the possibility of contamination with non-nucleotide ^{32}P excluded, these results, which appear to be contradictory to those of Marshak & Vogel (1950) and of Smellie *et al.* (1953) with alkaline digests of RNA, cannot be considered definitive. Logan & Smellie (1956) have conducted experiments in which isolated nuclei were incubated with cytoplasmic constituents, either component having previously been labelled *in vivo* with ^{32}P. Little ^{32}P appeared to pass from the nuclei to the cytoplasmic RNA although ^{32}P-labelled cytoplasm appeared to contribute a considerable amount of ^{32}P to both the DNA and RNA of the nuclei incubated in it. They conclude that their results disprove the hypothesis that the nuclear RNA may be the precursor of the cytoplasmic RNA. In the writer's opinion, the ^{32}P appearing in the nuclear fractions was introduced as a contaminant by precipitation and adsorption of cytoplasmic constituents on to the nuclei by the acid used in their subsequent isolation. Schneider & Potter (1957) have found a difference in the behaviour of *in vivo*-, as compared with *in vitro*-labelled microsomes, in that the former do not exhibit the reduction in specific activity on incubation shown by the latter. Paterson & Le Page (1957) have presented evidence that *in vitro* RNA 'renewal' may be the result of the addition of nucleotides to the 5′-ends of the existing molecules rather than by general replacement. These results suggest the possibility that sampling of RNA nucleotides by incomplete degradative procedures may yield fractions which are not representative of any population of RNA molecules, such fractions containing greater or lesser amounts of terminal nucleotides. The observation by Paterson & Le Page that renewal of RNA in homogenates is enhanced by the addition of damaged nuclei and nucleotide diphosphates is of interest in connexion with the nuclear RNA precursor hypothesis. Recently it has been reported that, following $^{32}PO_4$ administration, two types of RNA may be extracted from thymus nuclei, which differ in the distribution of ^{32}P in their constituent nucleotides. However, both types had specific activities much greater than the *c*-RNA (Logan & Davidson, 1957).

From the observation that a tissue whose cells were in active mitosis incorporated ^{32}P into the *n*-RNA for some time before any appeared in the DNA, it was postulated that the *n*-RNA might not only function as a precursor of *c*-RNA, but under appropriate circumstances might also be a precursor of DNA (Marshak, 1948). Subsequently, a thymine-containing compound was found in the acid-soluble fraction of *Arbacia* eggs and shortly thereafter deoxyribonucleotides were found in very small but significant amounts in liver; the interesting observation was made that there was an apparent increase in the deoxyribonucleotide level during liver regeneration (Marshak, 1954; Marshak & Marshak 1955 *b*; Potter & Schlesinger, 1955; Schneider, 1955; Schneider & Brownell, 1956). These findings introduced the possibility that DNA was formed from small-molecule precursors, rather than by the route just mentioned. However, in experiments in which orotic acid-^{14}C was administered to rats, the specific activity-time relations for the acid-soluble constituents, *n*-RNA, and DNA, of the regenerating liver were such that the deoxyribonucleotides did not appear to be DNA precursors, although it did seem possible that either the free ribonucleotides or those of RNA might function in this way (Hecht & Potter, 1956). When slices of regenerating liver were incubated with labelled orotic acid, it was found that *n*-RNA became labelled immediately, but that *c*-RNA and DNA became labelled only after an initial lag (Hecht *et al.* 1957). This observation appears to be analogous to that reported for ^{32}P *in vivo* by Marshak (1948) and supports the hypothesis that *n*-RNA may function as the precursor of DNA.

In the case of the *E. coli* bacteriophages, the newly synthesized phage formed after infection appeared to derive most of the bases for their DNA from the DNA of the bacterial host (Hershey, Garen, Fraser & Hudis, 1954). However, the possibility could not be excluded that bacterial RNA might contribute some of the bases, or else might serve some intermediary function (Hershey, 1953, 1956). Recent experiments by Volkin & Astrachan (1956), Astrachan & Volkin (1957) with the phage T2 have given results resembling those reported for the n-RNA, c-RNA, DNA relationship in mammalian tissues. After infection two particulate fractions were obtained, one of which had RNA with at least fifteen times greater specific activity than the other. The RNA which had the high turnover rate was only a small portion of the total bacterial RNA, a situation analogous to the relative proportions of n-RNA and c-RNA in mammalian tissues. When infected cells were exposed to $^{32}PO_4$, incorporation into RNA was initially more rapid than into DNA, again resembling the mammalian n-RNA, DNA relation. Chloramphenicol inhibited both ^{32}P release from RNA and synthesis of DNA, while removal of the chloramphenicol resulted in concomitant release of ^{32}P from RNA and its incorporation into DNA. While DNA was being synthesized and ^{32}P incorporated into it, only the active RNA fraction lost ^{32}P. The pattern assumed by the quantities of ^{32}P accumulated by each of the nucleotides of the active RNA resembled that of the proportions of bases of the DNA to be synthesized. From their more recent studies on chloramphenicol-inhibited phage, Hershey & Melechen (1957) infer that a phage precursor, DNA-1, is formed from some undefined intermediates and from bacterial DNA; DNA-1 in turn gives rise to a second DNA, which is combined with a sulphur-containing protein, and this is the immediate phage precursor. It seems likely that an active RNA precursor may intervene in this sequence as suggested by the preceding observations. This lends further support to the hypothesis that there is a kind of RNA, such as that found in the cell nucleus, which plays an active role in the synthesis of DNA. Chloramphenicol inhibits protein synthesis as well as RNA turnover, which leaves open the question of whether not only RNA, but also protein turnover may be involved in the laying down of new DNA. Other evidence for the probable intervention or association of protein synthesis with the laying down of new DNA is to be found in experiments with inhibitors of protein synthesis by Burton (1955), Melechen & Hershey (Melechen, 1955), and by Tomizawa & Sunakawa (1956).

The experiments mentioned above raise the question whether or not the DNA is an adequate and competent genetic agent. I have mentioned earlier the failure of this concept to account for the disappearance of DNA during the maturation of echinoderm eggs. If, in order to function genetically, DNA must operate through the activation of or with the participation of RNA or protein, serious doubt is introduced as to the validity of the concept that the gene is nothing but DNA, for the observations with phage suggest the probable intervention of RNA, protein, or both, as replica-forming material laid down between the time of introduction of the parent phage and the appearance of progeny DNA. When taken together with the existence of viruses such as tobacco mosaic, whose genetic material seems to be RNA or RNA-protein (Fraenkel-Conrat & Williams, 1955; Fraenkel-Conrat, Singer & Williams, 1957; Gierer & Schramm, 1956), the concept of a single substance as the gene is difficult to maintain, in spite of the wealth of evidence supporting the proposition that some kind of genetic function is fulfilled by either or both nucleic acids. In a previous paper, two series of studies, one on nucleic acid turnover, and a second analysing the response of chromosomes to ionizing radiation were reviewed and the conclusion reached that at least three kinds of substances were involved in gene and chromosome replication, namely RNA, DNA, and histone (Marshak, 1955). A similar conclusion is reached from the experiments reviewed above. In our concentration on one or the other of the nucleic acids and their potential information-transmitting capacity (Gamow, 1954a, b, 1955), we have been inclined to overlook the fact that in the reproducing system itself, the nucleic acids seem to exist as nucleo-proteins. With phage and tobacco mosaic virus, it is possible that the nucleic acids alone may be the infective agents, but we still lack critical evidence that, once inside the host cell, they function as the free acids rather than as nucleoproteins.

In these and in all previous discussions on the role of the nucleic acids in autosynthesis or in the biosynthesis of other macromolecules, the role of these substances has been dealt with entirely in terms of the chemical reactions they may participate in, with or without the assistance of enzymes or other adjuvants. Looked at in this way, this role is a passive one, since these compounds are imagined to determine the direction and velocity of intracellular chemical reactions solely by virtue of their chemical properties as ingredients in the reaction mixture. Even regarded as templates, equipped as in the Watson & Crick model, with a particular spatial configuration, we can do no more with such models than find loci for particular chemical affinities; this of course is a traditional point of view. The shapes of intra-cellular structures are sometimes given some attention, but usually only in relation to possible chemical affinities or to permeability of the structures' surfaces.

Are we, then, to consider the cell as a sort of miniscule reaction vessel, equipped it is true, with pumps and pores for regulating input and exit of materials and with some incidental devices for mixing and sorting, but one wherein events and products are determined solely by chemical interactions of its inherited substances with ingredients introduced from the environment? In trying to understand the nature of genic control by the use of this sort of concept we are led to an obvious conflict: genes of one kind must be DNA; genes of another must be RNA. Similarly, a cell without a nucleus may make RNA for a while but is doomed to ultimate extinction; a cell with a nucleus but apparently lacking DNA will go on to multitudinous duplications generating DNA as well as RNA. In seeking an explanation for this discordance, biochemistry alone seems inadequate and it appears necessary to look into the physical properties of chromosomes and other cell structures to seek attributes which may be characteristic of the assemblage but which are not necessarily possessed by the units of which it is composed.

The helical chromonemata seen on examination of chromosomes with the light microscope are now known to be composed of hierarchies of helical sub-units. In *E. coli* Marshak (1951*c*, *d*) found at least eight substituent helices per chromosome, the smallest being 100–150 Å. in diameter. In *Tradescantia*, Kaufmann & MacDonald (1956) found at least sixty-four sub-units, the diameter of the smallest being 125 Å. in diameter. Pairing of helices is obvious in some of the hierarchies but is by no means clear in all of them. It has been suggested, though without direct supporting evidence, that the lower-order helices may be associated lengthwise in groups to form ribbons (Taylor, Woods & Hughes, 1957). In cross-section the 125 Å. helix appeared to contain a dense wall and a less dense core. The wall was about 40 Å. thick and the diameter of the core was about 40 Å. (Kaufmann & MacDonald, 1956). MacDonald & Kaufmann have suggested that the wall may be the ultimate pair of sub-units and might be analogous to the paired nucleoprotein fibrils of the Wilkins model for deoxyribonucleoprotein (Wilkins, 1956). In cross-section the smaller fibrils have an appearance which resembles the tobacco mosaic virus 'doughnuts', and it has been suggested that this fibril has a central core of nucleic acid which is not osmophilic and therefore appears less dense (Bahr, 1954; Ris, 1956). Although the work of Fraenkel-Conrat & Williams suggested a nucleic acid core surrounded by a protein helix as the structure for tobacco mosaic virus, subsequent studies by X-ray scattering methods applied to the virus and to the protein component alone indicated the presence of a hollow core, about which was wound the helical RNA chain embedded in protein (Caspar, 1956; Franklin, 1956).

Marshak & Takahashi (1942) showed that, in tobacco mosaic virus irradiated with X-rays, energy absorbed in one member of a virus polymer could be transferred to other members. This transfer occurred when the virus elements were linked by chemical bonds, but not when they were aggregated by electrostatic forces. Siegel and his co-workers have found a marked difference in the ultra-violet inactivation curves of two strains of tobacco mosaic virus and presented evidence that this was due to differences in the way the protein was bound to the nucleic acid. Their results indicated conduction of absorbed energy from the nucleic acid to the protein moiety (Siegel & Wildman, 1954, 1956; Siegel, Wildman & Ginoza, 1956). If we accept such conduction as experimentally established in tobacco mosaic virus it is not unreasonable to expect it also to occur in chromosomes.

Doty and Rice have found that the hydrogen bonds which hold together the bases of the 2-stranded DNA helix may be ruptured by relatively moderate elevation of the temperature (Doty & Rice, 1955; Thomas & Doty, 1956). By subjecting polyribonucleotide aggregates to elevated temperature and urea, Warner (1957) found dissociation of the aggregates with accompanying increase in ultra-violet absorption, which was interpreted as the result of breakage of hydrogen bonds linking the bases. These results with chemical systems have their counterpart in observations on the nuclei of living cells. About ten years ago it was reported that undamaged nuclei showed no specific absorption at 260 mμ although such absorption became pronounced following damage due to exposure to ultra-violet (Brumberg & Larionow, 1946). Essentially the same observation has been made more recently with the ultra-violet flying-spot microscope (Montgomery, Bonner & Roberts, 1956). When taken together with the results obtained with the chemical systems, the observations suggest that the increase in ultra-violet absorption by the nucleus may be due to breakage of bonds resulting in the equivalent of a 'hyperchromic effect' in the chromosomes. Lawley (1956) has presented an adequate explanation for the increased absorption at 260 mμ found when DNA is denatured. He postulated that in the transition from the undenatured to the denatured state there was an increase in the effective chromophore area brought about by randomization of the ordered stacking of the bases. This, however, is not a necessary interpretation and another related to the postulated semi-conductor property of nucleic acids and chromosomes also seems applicable. Privat de Garilhe & Laskowski (1956) have observed the hyperchromic effect in the cleavage of dinucleotides and found a marked difference in the amount of hyperchromaticity in pyrimidine dinucleotides as compared with those having one or two purines. Their observations do not appear to be interpretable in terms of Lawley's effective-chromophore-area hypothesis.

Assuming that the chromosome fibril does act as a semi-conductor, an electron, or electron vacancy, may travel within it for short distances, for example to the nearest —S—S group as has been suggested by Gordy for proteins irradiated with X-rays (Gordy, Ard & Shields, 1955). If the pairs of fibrils are connected at their distal ends as well as at the kinetochore a continuous circuit would be established. Current flow would generate a magnetic field which in turn could impose a torque upon other current-carrying helical elements of the higher order about which the smaller elements are wound. Forces such as these may be the ones which are effective in separating parent from daughter helices which must be rotated about each other in order to disjoin, and by acting through the hierarchy of the various orders of helices result in the separation of the microscopically visible half-chromatids. Since separation of parent from daughter strands occurs only once in the life history of the cell, the mechanism just described may be operative only during a single stage. The short-range conductions between low-energy electron traps such as the —S—S bonds may operate independently of the chromosome reproduction process. There may, therefore, very well be short regions of the chromosome fibrils which act as electromagnetic oscillators each with a characteristic period determined by the properties of the particular coupled RNA, DNA, and protein of that region. All of these units may be coupled into an oscillating circuit comprising the whole chromosome, and, by way of links through the nucleolus organizers and the chromocentres, encompassing the whole chromosome complement. A source of power for driving these oscillating systems might be derived from energy input at nucleotide turnover. If each nucleotide exchange was accompanied by a free radical derived from the nucleotide di- or triphosphate to monophosphate conversion, it could serve to deliver a pulse to the system. If the period of the overall oscillating system, i.e. the chromosome circuit, is determined solely by the rate of input of such pulses, this rate as calculated from RNA turnover data is about 10^6 nucleotides per second per rat liver nucleus. The wavelength associated with this frequency is much too long to have a significant cellular effect. However, the period of oscillation will be determined not only by the input rate but by other characteristics of the oscillating systems, and it seems not unreasonable to expect wavelengths in an intracellularly effective range. In this connexion it is worth noting that electron spin intercombinations are known to occur in molecules composed entirely of low —Z atoms, but such coupling is greatly increased in molecules in which a high —Z atom is linked to low —Z atoms (Kasha & McGlynn, 1956) thus giving a greater probability of frequency transitions in such molecules. The nucleic acids are linked series of such combinations and may therefore be expected to facilitate frequency transitions and thus have a coupling as well as a conducting function.

If we accept this hypothesis, the helical shape of the chromosome sub-units and of DNA and proteins has a physical significance in the cell economy beyond that of being the accidental result of the association of the kind of atoms needed in cell metabolism and synthesis. They provide a structural configuration suited to their function as oscillating systems as well as units of reproduction. The shapes of components of the cytoplasm would also have meaning in the light of this consideration. For example, the vesicular character of the endoplasmic reticulum with its regularly spaced microsomes, suggests a role as a transducer of electromagnetic radiation from the nucleus. The mitochondrion with its chambers and cristae suggests a resonant cavity with wave channels. The 'pores' of the nuclear membrane, which are much too large for any permeability function, may have significance in this connexion. Similarly the changes in amount of heterochromatin with differentiation and with alteration in environmental conditions may have a relation to the conducting and emitting properties of the chromosomes. All of this may seem not only speculative but fanciful. Nevertheless something of this sort is needed to account for what we see in cells. The particular shapes of intracellular structures repeated in their essential characteristics throughout a great variety of cells are meaningless on a purely chemical scheme, and their existence would be difficult to account for by evolution unless related to a direct controlling influence by the nucleus. The theory provides a means for understanding some genetic phenomena which are otherwise not readily explained. The term gene may be taken to represent a localized region of the chromosome which has an activity characteristic of that region (Beadle, 1957; Marshak, 1951 b). Each such region may then be identified with the localized oscillators of the chromosome in our scheme. The phenomenon of hetero-allelism (Roman, 1956) may then be attributed to the association of two alleles each with a somewhat different chemical structure and hence a slightly different mode of oscillation, the unbalanced condition inducing further mutation. The Ds-Ac 'modifier' system of Zea (McClintock, 1956) then becomes one in which the Ds and Ac units introduce into the overall oscillating circuit elements equivalent to altered inductance or capacity, which would have an influence on the oscillation behaviour of chromosome regions and of individual genes.

These notions have little value unless they stimulate experimentation. They are presented because I believe that in both biochemistry and genetics, we are in need of new avenues of approach to long-standing problems, and because the ideas presented here can be subjected to experimental test.

REFERENCES

ALLFREY, V., MIRSKY, A. E. & OSAWA, W. (1957). In *The Chemical Basis of Heredity*, p. 200. Ed. McElroy, W. D. & Glass, B. Baltimore: Johns Hopkins Press.

ASTRACHAN, L. & VOLKIN, E. (1957). *Fed. Proc.* **16**, 147.

BAAS-BECKING, L. G. M., BAKHUYSEN, H. V. D. S. & HOTELLING, H. (1928). *Verh. Akad. Wet. Amst.* **25**, 5.

BAHR, G. F. (1954). *Exp. Cell Res.* **7**, 457.

BARNUM, C. P. & HUSEBY, R. A. (1950). *Arch. Biochem.* **29**, 7.

BARNUM, C. P., HUSEBY, R. A. & VERMUND, H. (1952). *Cancer Res.* **12**, 246.

BARNUM, C. P., HUSEBY, R. A. & VERMUND, H. (1953). *Cancer Res.* **13**, 880.

BEADLE, G. (1957). In *The Chemical Basis of Heredity*, p. 3. Ed. McElroy, W. D. & Glass, B. Baltimore: Johns Hopkins Press.

BENDICH, A., PAHL, H. B. & BROWN, G. B. (1957). In *The Chemical Basis of Heredity*, p. 378. Ed. McElroy, W. D. & Glass, B. Baltimore: Johns Hopkins Press.

BRACHET, J. & CHANTRENNE, H. (1956). *Cold Spr. Harb. Symp. Quant. Biol.* **21**, 329.

BRUMBERG, E. M. & LARIONOW, L. T. (1946). *Nature, Lond.* **158**, 663.

BUCHANAN, J. M. (1952). In *Phosphorus Metabolism*. Ed. McElroy, W. D. & Glass, B., **2**, p. 406. Baltimore: Johns Hopkins Press.

BURTON, K. (1955). *Biochem. J.* **61**, 473.

CASPAR, D. L. D. (1956). *Nature, Lond.* **177**, 928.

COHN, W. (1956). In *Currents in Biochemical Research*, p. 460. Ed. Green, D. V. New York: Interscience Publishing Co.

CRICK, F. H. C. (1957). In *The Chemical Basis of Heredity*, p. 532. Ed. McElroy, W. D. & Glass, B. Baltimore: Johns Hopkins Press.

CRICK, F. H. C. & WATSON, J. D. (1954). *Proc. Roy. Soc.* A, **223**, 80.

DOTY, P. & RICE, A. S. (1955). *Biochim. Biophys. Acta*, **16**, 446.

DUNN, D. B. & SMITH, J. D. (1955). *Nature, Lond.* **175**, 336.

ELSON, D., TRENT, L. T. W. & CHARGAFF, E. (1955). *Biochim. Biophys. Acta*, **17**, 362.

FRAENKEL-CONRAT, H., SINGER, B. A. & WILLIAMS, R. C. (1957). In *The Chemical Aspects of Heredity*, p. 501. Ed. McElroy, W. D. & Glass, B. Baltimore: Johns Hopkins Press.

FRAENKEL-CONRAT, H. & WILLIAMS, R. C. (1955). *Proc. Nat. Acad. Sci., Wash.* **41**, 690.

FRANKLIN, R. (1956). *Nature, Lond.* **177**, 928.

FRESCO, J. & MARSHAK, A. (1954). *J. Biol. Chem.* **205**, 585.

GALE, E. F. & FOLKES, J. P. (1955a). *Biochem. J.* **59**, 661, 675.

GALE, E. F. & FOLKES, J. P. (1955b). *Nature, Lond.* **175**, 592.

GAMOW, G. (1954a). *Nature, Lond.* **173**, 318.

GAMOW, G. (1954b). *Biol. Medd. Kgl. Danske Vidensk. Selsk.* **22**, 3.

GAMOW, G. (1955). *Proc. Nat. Acad. Sci., Wash.* **41**, 7.

GAY, H. (1956). *Cold. Spr. Harb. Symp. Quant. Biol.* **21**, 257.

GIERER, A. & SCHRAMM, G. (1956). *Nature, Lond.* **177**, 702.

GOLDSTEIN, L. & PLAUT, W. (1955). *Proc. Nat. Acad. Sci., Wash.* **41**, 871.

GOLDTHWAIT, D. A., GREENBERG, G. R. & PEABODY, R. A. (1955). *Biochim. Biophys. Acta*, **18**, 148.

GOLDWASSER, E. (1955). *J. Amer. Chem. Soc.* **77**, 6083.

GORDY, W., ARD, W. B. & SHIELDS, H. (1955). *Proc. Nat. Acad. Sci., Wash.* **41**, 983, 996.

GOTS, J. S. & GOLLUB, E. G. (1957). *Proc. Amer. Ass. Cancer Res.* **2**, 207.

GREENBERG, G. R. & SPILLMAN, E. L. (1956). *J. Biol. Chem.* **219**, 411.

GRÜNBERG-MANAGO, M. & OCHOA, S. (1955). *J. Amer. Chem. Soc.* **77**, 3165.

GRÜNBERG-MANAGO, M., ORITZ, P. J. & OCHOA, S. (1955). *Science*, **122**, 907.
GRÜNBERG-MANAGO, M., ORITZ, P. J. & OCHOA, S. (1956). *Biochim. Biophys. Acta*, **20**, 269.
HARTMAN, S. C., LEVENBERG, B. & BUCHANAN, J. M. (1956). *J. Biol. Chem.* **221**, 1057.
HECHT, L. F., BRUMM, A. F. & POTTER, V. R. (1957). *Proc. Amer. Ass. Cancer Res.* **2**, 212.
HECHT, L. & POTTER, V. R. (1956). *Cancer Res.* **16**, 988, 999.
HECHT, L., POTTER, V. R. & HERBERT, E. (1954). *Biochim. Biophys. Acta*, **15**, 134.
HERSHEY, A. D. (1953). *J. Gen. Physiol.* **37**, 1.
HERSHEY, A. D. (1956). In *Currents in Biochemical Research*, p. 1. Ed. Green, D. E. New York: Interscience Publishing Co.
HERSHEY, A. D., GAREN, A., FRASER, D. & HUDIS, J. D. (1954). *Yearb. Carneg. Instn*, **53**, 210.
HERSHEY, A. D. & MELECHEN, N. E. (1957). *Virology*, **3**, 207.
HOAGLAND, J. C. (1955). *Biochim. Biophys. Acta*, **16**, 288.
HOAGLAND, J. C. & ZAMECNIK, P. C. (1957). *Fed. Proc.* **16**, 197.
HÖBER, R. (1945). In *Physical Chemistry of Cells and Tissues*, ch. 10. Philadelphia: The Blakiston Co.
HOROWITZ, N. H. & FLING, M. (1956). In *Enzymes: Units of Biological Structure and Function*, p. 139. Ed. Gaebler, O. H. New York: Academic Press.
HURLBERT, R. B. & POTTER, V. R. (1952). *J. Biol. Chem.* **195**, 257.
HURLBERT, R. B., SCHMITZ, H., BRUMM, A. F. & POTTER, V. R. (1954). *J. Biol. Chem.* **209**, 23, 41.
IRVIN, J. L., ROTHERHAM, J., IRVIN, E. M. & HOLBROOK, D. J. (1957). *Fed. Proc.* **16**, 199.
JACOB, F. & WOLLMAN, E. L. (1957). In *The Chemical Basis of Heredity*, p. 468. Ed. McElroy, W. D. & Glass, B. Baltimore: Johns Hopkins Press.
JEENER, R. & SZAFARZ, D. (1950). *Arch. Biochem.* **26**, 54.
KACSER, H. (1956). *Science*, **124**, 151.
KASHA, M. & McGLYNN, S. P. (1956). *Annu. Rev. Phys. Chem.* **7**, 403.
KAUFMANN, B. P. & MACDONALD, M. R. (1956). *Cold Spr. Harb. Symp. Quant. Biol.* **21**, 233.
KELLER, E. B., ZAMECNIK, P. C. & LOFTFIELD, R. B. (1954). *J. Histochem. Cytochem.* **2**, 378.
KORNBERG, A. (1957). In *The Chemical Basis of Heredity*, p. 579. Ed. McElroy, W. D. & Glass, B. Baltimore: Johns Hopkins Press.
KROGH, A. (1946). *Proc. Roy. Soc.* B, **133**, 140.
KRUPKO, S. & DENLEY, A. (1956). *Nature, Lond.* **177**, 92.
LAWLEY, P. D. (1956). *Biochim. Biophys. Acta*, **21**, 481.
LEHNINGER, A. (1956). In *Enzymes: Units of Biological Structure and Function*. Ed. Gaebler, O. H. New York: Academic Press.
LEVENBERG, B. & BUCHANAN, J. M. (1956). *J. Amer. Chem. Soc.* **78**, 504.
LEVINTHAL, C. (1956). *Proc. Nat. Acad. Sci., Wash.* **42**, 394.
LINDBERG, O. & ERNSTER, L. (1954). 'Chemistry and Physiology of Mitochondria and Microsomes'. In *Handbuch der Protoplasmsforschung*. Ed. Heilbrunn, L. V. & Weber, F. Vol. 3. London: Lange, Maxwell and Springer, Ltd.
LOGAN, R. & DAVIDSON, J. D. (1957). *Biochim. Biophys. Acta*, **24**, 196.
LOGAN, R. & SMELLIE, R. M. S. (1956). *Biochim. Biophys. Acta*, **21**, 92.
LUDWIG, K. S. (1954). *Arch. Biol.* **65**, 137.
McCLINTOCK, B. (1956). *Cold Spr. Harb. Symp. Quant. Biol.* **21**, 197.
MAGASARIK, B. (1955). In *The Nucleic Acids*, p. 393. Ed. Chargaff, E. & Davidson, J. N. Vol. 2. New York: Academic Press.

MARSHAK, A. (1941). *J. Gen. Physiol.* **25**, 275.
MARSHAK, A. (1948). *J. Cell. Comp. Physiol.* **32**, 381.
MARSHAK, A. (1951a). *J. Biol. Chem.* **189**, 607.
MARSHAK, A. (1951b). *Cold Spr. Harb. Symp. Quant. Biol.* **16**, 156.
MARSHAK, A. (1951c). *Proc. Nat. Acad. Sci., Wash.* **37**, 38.
MARSHAK, A. (1951d). *Exp. Cell Res.* **2**, 243.
MARSHAK, A. (1954). *Biochim. Biophys. Acta,* **15**, 584.
MARSHAK, A. (1955). *Trans. N.Y. Acad. Sci.* **17**, 506.
MARSHAK, A. (1957). *Exp. Cell Res.* (in the Press).
MARSHAK, A. & CALVET, F. (1949). *J. Cell. Comp. Physiol.* **34**, 451.
MARSHAK, A. & FAGER, J. (1950). *J. Cell. Comp. Physiol.* **35**, 317.
MARSHAK, A. & MARSHAK, C. (1953). *Exp. Cell Res.* **5**, 288.
MARSHAK, A. & MARSHAK, C. (1955a). *Exp. Cell Res.* **8**, 126.
MARSHAK, A. & MARSHAK, C. (1955b). *J. Biophys. Biochem. Cytol.* **1**, 167.
MARSHAK, A. & TAKAHASHI, W. N. (1942). *Proc. Nat. Acad. Sci., Wash.* **28**, 211.
MARSHAK, A. & VOGEL, H. (1950). *J. Cell. Comp. Physiol.* **36**, 97.
MARSHAK, A. & WALKER, A. C. (1945). *Amer. J. Physiol.* **143**, 235.
MAZIA, D. (1956). In *Enzymes: Units of Biological Structure and Function,* p. 261. Ed. Gaebler, O. H. New York: Academic Press.
MELECHEN, N. E. (1955). *Genetics,* **40**, 585.
MIRSKY, A. E., OSAWA, S. & ALLFREY, V. G. (1956). *Cold Spr. Harb. Symp. Quant. Biol.* **21**, 49.
MOLDAVE, K. & HEIDELBERGER, C. (1954). *J. Amer. Chem. Soc.* **76**, 679.
MONTGOMERY, P. O'B., BONNER, W. A. & ROBERTS, F. F. (1956). *Proc. Soc. Exp. Biol., N.Y.* **93**, 409.
OCHOA, S. & HEPPEL, L. A. (1957). In *The Chemical Basis of Heredity,* p. 615. Ed. McElroy, W. D. & Glass, B. Baltimore: Johns Hopkins Press.
PARK, J. T. (1952). *J. Biol. Chem.* **194**, 877, 885, 897.
PATERSON, A. R. P. & LE PAGE, G. A. (1957). *Cancer Res.* **17**, 409.
POTTER, R. L., HECHT, L. & HERBERT, E. (1956). *Biochim. Biophys. Acta,* **20**, 439.
POTTER, R. L. & SCHLESINGER, S. (1955). *J. Amer. Chem. Soc.* **77**, 6714.
POTTER, R. L., SCHNEIDER, J. H. & HECHT, L. I. (1957). In *The Chemical Basis of Heredity,* p. 639. Ed. McElroy, W. D. & Glass, B. Baltimore: Johns Hopkins Press.
PRESCOTT, D. M. (1957). *Exp. Cell. Res.* **12**, 196.
PRIVAT DE GARILHE, M. & LASKOWSKI, M. (1956). *J. Biol. Chem.* **223**, 661.
REMY, C., REMY, W. T. & BUCHANAN, J. M. (1955). *J. Biol. Chem.* **217**, 885.
RIS, H. (1956). In *The Chemical Basis of Heredity,* p. 23. Ed. McElroy, W. D. & Glass, B. Baltimore: Johns Hopkins Press.
ROLL, P. M., WEINFELD, H., CARROLL, E. & BROWN, G. B. (1956a). *J. Biol. Chem,* **220**, 439.
ROLL, P. M., WEINFELD, H. & CARROLL, E. (1956b). *J. Biol. Chem.* **220**, 454.
ROMAN, H. L. (1956). *Cold Spr. Harb. Symp. Quant. Biol.* **21**, 175.
ROSE, I. A. & SCHWEIGERT, B. S. (1953). *J. Biol. Chem.* **202**, 635.
SABLE, H. Z., WILBER, P. B., COHEN, A. E. & KANE, M. R. (1954). *Biochim. Biophys. Acta,* **13**, 156.
SACKS, J. & SAMARTH, K. D. (1956). *J. Biol. Chem.* **223**, 423.
SCHMITZ, H., HURLBERT, R. B. & POTTER, V. R. (1954). *J. Biol. Chem.* **209**, 41.
SCHNEIDER, J. H. & POTTER, V. R. (1957). *Fed. Proc.* **16**, 243.
SCHNEIDER, W. C. (1955). *J. Biol. Chem.* **216**, 287.
SCHNEIDER, W. C. & BROWNELL, L. W. (1956). *Fed. Proc.* **15**, 349.
SEIFRIZ, W. (1942). In *The Structure of Protoplasm.* Ames, Iowa: Iowa State College Press.

SIEGEL, A. & WILDMAN, S. G. (1954). *Phytopathology*, **44**, 277.

SIEGEL, A. & WILDMAN, S. G. (1956). *Virology*, **2**, 69.

SIEGEL, A., WILDMAN, S. G. & GINOZA, W. (1956). *Nature, Lond.* **178**, 1117.

SMELLIE, R. M. S. (1955). In *The Nucleic Acids*, p. 408. Ed. Chargaff, E. & Davidson, J. N. Vol. 2. New York: Academic Press.

SMELLIE, R. M. S. & DAVIDSON, J. N. (1956). *Experientia*, **12**, 422.

SMELLIE, R. M. S., McINDOE, W. M. & DAVIDSON, J. N. (1953). *Biochim. Biophys. Acta*, **11**, 559.

SMELLIE, R. M. S., McINDOE, W. M., LOGAN, R., DAVIDSON, J. N. & DAWSON, I. M. (1953). *Biochem. J.* **54**, 280.

SPEK, J. (1930). *Protoplasma*, **9**, 370.

SPEK, J. (1934). *Protoplasma*, **21**, 394.

SPEK, J. & CHAMBERS, R. (1933). *Protoplasma*, **20**, 376.

SPIEGELMAN, S. (1957). In *The Chemical Basis of Heredity*, p. 232. Ed. McElroy, W. D. & Glass, B. Baltimore: Johns Hopkins Press.

TAYLOR, J. H., WOODS, P. S. & HUGHES, W. L. (1957). *Proc. Nat. Acad. Sci., Wash.* **43**, 122.

THOMAS, C. A., JR. & DOTY, P. (1956). *J. Amer. Chem. Soc.* **78**, 1854.

TOMIZAWA, J. & SUNAKAWA, S. (1956). *J. Gen. Physiol.* **39**, 553.

TYNER, E. P., HEIDELBERGER, C. & LE PAGE, G. A. (1953). *Cancer Res.* **13**, 186.

VOLKIN, E. & ASTRACHAN, L. (1956). *Virology*, **2**, 149.

VON BORSTEL, R. C. (1955). *Nature, Lond.* **175**, 342.

WARNER, R. C. (1957). *Fed. Proc.* **16**, 266.

WATANABE, I. & WILLIAMS, C. W. (1953). *J. Gen. Physiol.* **37**, 71.

WATSON, J. D. & CRICK, F. H. C. (1953). *Nature, Lond.* **171**, 737.

WILKINS, M. H. F. (1956). *Cold Spr. Harb. Symp. Quant. Biol.* **21**, 75.

WILKINS, M. H. F., STOKES, A. R. & WILSON, H. R. (1953). *Nature, Lond.* **171**, 738.

WILLIAMS, W. L. & BUCHANAN, J. M. (1953). *J. Biol. Chem.* **203**, 583.

YASUYUKI, T., HECHT, L. I. & POTTER, V. R. (1956). *Cancer Res.* **16**, 994.

ZAMECNIK, P. C. & KELLER, E. B. (1954). *J. Biol. Chem.* **209**, 337.

TISSUE TRANSPLANTATION AND CELLULAR HEREDITY

By N. A. MITCHISON

Zoology Department, University of Edinburgh

I. INTRODUCTION

Little is known about the determinants of cellular heredity in somatic cells. Now that so much is known about the mechanism of heredity in unicellular organisms, it seems reasonable to seek for similar information about cells in tissues. Although it must be admitted that the information which has been obtained so far is patchy, there is general agreement about the sort of questions which can usefully be asked. A rough sequence might run as follows. Which cell characters are sufficiently precise and easily detectable to be useful as markers in genetic studies? Which markers are determined by the cellular environment, and which by heredity? How do the determinants alter during normal embryonic differentiation, and can they be made to mutate under controlled conditions? Can mutations be selected? Can the determinants pass from one cell to another, thus bringing about genetic recombination? Are the determinants the nuclear genes themselves, and if not, what is their relationship to the genes? What is the physical and chemical nature of the determinants? A list of this sort is bound to be partly a reflexion of personal opinion: a taste for old-fashioned genetics is expressed by putting the last question last. More extensive discussions of these questions may be found in recent discussions by Ephrussi (1956), Lederberg (1956), Medawar (1947), Sonneborn (1950) and Trinkaus (1956).

It is commonly held that two requirements must be met, in order that somatic genetics can be understood at all. The first is that tissue cells should be undemocratic. Unless there is some privileged and restricted class of molecule or particle which tells the rest of the cell what to do, the prospect seems hopeless. But there is no reason for pessimism, when such strong candidates as the nuclear genes, steady-state key enzymes, and RNA-containing cytoplasmic particles are in the field. The nuclear genes themselves have usually been excluded for various reasons, but the outcome of nuclear transplantation in Amphibia has shown that this may have been premature (Briggs & King, 1956).

The second requirement is that genetic recombination should occasionally take place. No doubt the accumulation of markers, and sieves for their selection, is especially important for their use in detecting recombination. But at the moment the examples of recombination in tissue cells are uncertain and isolated. Nevertheless, the accumulation of markers may well turn out to be worth while, for their use in mutational analysis. Mutation is used here in the broadest sense of the word, to denote any sudden and clear-cut change in a hereditary character, without special reference to the mechanism of change. Quite apart from the possibility of recombination, mutations can be used to analyse control mechanisms according to the following principle. If a cellular character B alters whenever another character A alters, but A does not always alter when B does, A can be assumed to control B.

Cellular heredity can only be studied when cells multiply in a neutral environment, and this restricts the material which can be used. The main ways of persuading cells to multiply are tissue culture and tissue transplantation. Each has advantages. Tissue culture is better adapted for certain special purposes, such as the genetics of nutritional requirements, and perhaps the more quantitative experiments will be made with clones of cells growing in tissue culture (Puck & Marcus, 1955). On the other hand, differentiated cells seem to preserve their minute morphological and biochemical peculiarities better upon transplantation. There is no precedent in tissue culture, for example, for the detailed classification of the hereditary characteristics of skin epithelia, which Billingham & Medawar have made by heteroplastic skin grafts (Billingham & Medawar, 1950).

The three types of tissue which have been the chief subjects of genetical study by means of transplantation are skin, tumours, and the complex of spleen, lymph nodes, and bone marrow. Cells from each of these tissues can be made to go through repeated divisions by transplantation under appropriate circumstances. By this means, the cellular characteristics which are under genetic control can be sorted out from those which are adaptive. Genetic control here means simply that the character remains unaltered, after the cell has been in a neutral environment, and divided sufficiently to dilute out any determinant which is not replicated. Assuming a determinant can be no smaller than a hydrogen atom, the number of divisions needed to establish genetic control in an average-sized cell is about fifty. Thus, multiplication of spleen and bone marrow cells after injection into irradiated mice has shown that the following characters are under genetic control: capacity to produce erythrocytes with specific agglutinins, transplantation antigens, and chromosome morphology (Lindsley, Odell & Tausche, 1955; Mitchison, 1956a; Ford, Hammerton, Barnes & Loutit, 1956). By the same test, production of a specific antibody is possibly under genetic control; while the capacity to produce erythro-cytes with a specific dependence of osmotic fragility on temperature is definitely non-genetic (Mitchison, 1957; T. Makinodan & N. G. Anderson, unpublished).

Tumour cells have certain advantages over normal cells for genetic studies. The chief one is that they multiply naturally, without having to be put under special conditions. Because of this, the kind of genetic variation they show is fairly well known. Some of the special techniques for the genetics of micro-organisms have been successfully applied to tumour cells, such as mass selection for genetic markers, single cell iso-lation, and the fluctuation test (reviewed by Klein, 1956). A beginning has even been made with a systematic search for genetic recombination. The scope of this article is confined mainly to a single aspect of tumour cell genetics, mutations of antigens. The general idea of this line of research is that the immunological response to foreign tissue transplanted within a species—the homograft reaction—can be used to select mutations of iso-antigens in tumour cells. This in turn should lead to an understanding of the mechanism of mutation, and so throw light on the nature of the genetic determinants in somatic cells.

II. THE IMMUNOLOGICAL BASIS FOR TUMOUR TRANSPLANTABILITY

Surgically adequate grafts which regress in incompatible hosts do so because of an immunological reaction on the part of the host against the antigens of the graft. This generalization is firmly established for trans-plants of normal tissue and of tumours, even though some details of the immunological mechanism are not yet fully understood. It is based mainly on the difference in behaviour of first and second transplants into the same individual, and was widely accepted long before antibodies were demon-strated serologically (Woglom, 1929). Second and subsequent transplants into an individual which has been immunized by a first graft exhibit the phenomenon of accelerated graft-breakdown. Thus most tumours in mice grow initially and then regress, upon their first transplantation into an incompatible host. Transplants made subsequently are killed without a period of initial growth.

The features of this reaction which are important in the selection of mutations can be illustrated by a particular example: the reaction of mice of the C57BR/a strain to subcutaneous grafts of an A strain sarcoma, Sal. This account is abstracted from the work of Mitchison (1955), Mitchison & Dube (1955), and Andreini, Drasher & Mitchison (1955), and is summarized

15-2

in Fig. 1. Upon first transplantation into this incompatible strain, the
tumour has a median survival time of 10 days. During this time it grows
from a small transplant, and attains a size of around 100 mg. If the host
has been previously immunized by transplantation of the tumour or other

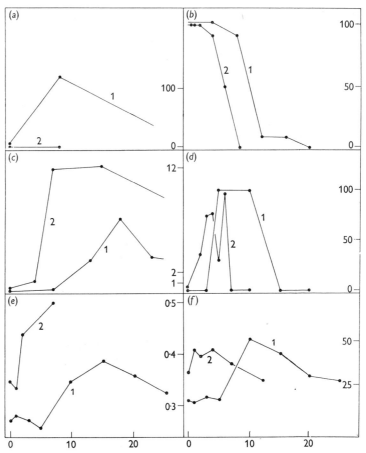

Fig. 1. The immunological response to foreign tumour transplants. Primary (1), and
secondary (2), transplants of sarcoma Sal in C57BR/a mice. The abscissa of every graph
is the interval after transplantation in days. (a) Tumour weight (mg.); (b) tumour survival
(% of original transplants); (c) agglutination titre of serum (log₂); (d) capacity of draining
lymph nodes to confer adoptive immunity (% of test grafts killed); (e) RNA/DNA ratio
in draining lymph nodes; (f) weight of draining lymph nodes (mg.).

A strain tissue, growth does not take place and the transplant is dead after
5 days. During the survival of the graft, antigens pass to the surrounding
lymph nodes, where they elicit cellular proliferation and the production of
antibodies. Between 3 and 5 days are needed for this process in normal
hosts, but it is accelerated in hosts which have been previously immunized.

During and after the destruction of the transplant, no effective antibody has been demonstrated in the serum by passive transfer tests, and it is surmised that this antibody is bound to the leucocytes of the blood. Nevertheless, an antibody directed against the antigens of the tumour can be detected in the serum, by the agglutination of A strain erythrocytes or the neutralization of tumour cells, according to the methods of Gorer & Mikulska (1954), and Gorer (1942).

The race which takes place between the growth of the tumour and the production of antibody in the lymph nodes is illustrated by an experiment with 'enhanced' mice. These were mice of the C57BR/a strain, whose own capacity to reject transplants of the tumour had been abolished by pre-treatment with lyophilized A strain tissue, according to a procedure of Kaliss & Snell (1951). The mice were implanted with the tumour, followed at various intervals by lymph nodes from donors immunized against the tumour. As the tumour increased in size, it was found to become progressively less susceptible to destruction by the lymph nodes.

Several features of the reaction can be exploited for mutation work. The initial growth of the tumour in non-immunized hosts provides a large population of cells in which mutations could occur, and the size of the population can be controlled by prior immunization. When the production of antibody commences, there is intense selection in favour of tumour cells which have mutated in their immunological properties, either towards loss of antigen, or to an increased ability to withstand immunological attack. Means of testing and characterizing mutations are provided by the specific serum antibodies, and also by the specific heightened graft-resistance of animals which have been immunized.

III. THE GENETICS OF THE TRANSPLANTATION ANTIGENS

A discussion of the genetics of the transplantation antigens is appropriate here for several reasons. First, genetical investigations have led to a theory of the relationship between the genes and these antigens. Any proposed mechanism of mutation must be consistent with this theory. In fact, the relationship between the genes and the antigens has turned out to be very simple, so that the possible mechanisms of mutations are limited. Secondly, individual antigens can be isolated because they are under the control of separate genes. Consequently, the identification of antigens has depended largely on genetical procedures. Thirdly, the genetical theory has been used to design and breed special mouse stocks for mutation work.

The genetical theory is based on the investigations of Little and his colleagues and successors at Bar Harbor, into the control of susceptibility to transplantable tumours in inbred mice (reviewed by Law, 1954). All the members of an inbred strain are normally susceptible to a tumour which originates in the strain, while foreign strains are not susceptible. Susceptibility here means that a subcutaneous transplant of the tumour is able to grow progressively and kill its host. In crosses between susceptible and non-susceptible strains, all the F_1 are susceptible, while most of the F_2 are not. A character thus inherited is one in which the simultaneous expression of a number of genes is necessary. The assumption was originally made that a tumour would grow progressively only in mice carrying certain dominant genes present in the strain of origin. As Snell (1948) has pointed out, a better assumption is that the genes concerned are semi-dominant: that is, they are active in the heterozygous as well as in the homozygous condition. This assumption serves to bring their genetic control into line with other systems of iso-antigens, such as the blood groups in man. The genetical theory was tested by the transplantation of various tumours into F_2 and back-crosses between susceptible and non-susceptible strains. In this way, independent estimates of the number of genes required for each tumour could be made, and were found to be in good agreement.

The genetical investigations showed that the genes appear to act independently of one another in the tumour, as well as in the host. This is shown by the transplantation of heterozygous tumours, originating in F_1 hybrids between two inbred strains. Thus Cloudman estimated the number of genes contributed separately by each parent in four of these tumours, by transplantation into each back-cross. The joint action of these genes was then determined, by transplantation into the F_2, and found to equal the sum of the two contributions in every case.

The inbred strains which were used in these experiments differed by a number of genes controlling susceptibility, ranging from one to about twelve. While most experiments were carried out with the A and DBA strain combination, there is enough evidence to show that the results with this strain combination are not exceptional. In order to be able to identify individual genes for susceptibility, Snell has undertaken an elaborate breeding programme (Snell, 1948). The procedure which has been used is analogous to the conventional method of creating isogenic strain by back-crossing. A susceptible strain, for example the C57BL strain, which grows a C57BL tumour, is crossed with a resistant strain, for example the DBA strain which resists the tumour. Mice of the F_2, F_4, F_6, etc. generations are inoculated with the C57BL tumour and the survivors are again mated to the susceptible strain. It can be calculated that after ten generations, the genotype of the isogenic resistant subline is 96·9% from the susceptible strain. In this way a series of genes for resistance has been isolated on the genetic background of susceptible strains. Eventually each isogenic strain is continued on its own by inbreeding.

The isogenic strains have been used by Snell and his colleagues to identify many of the genes controlling the transplantability of tumours. These genes are referred to as histocompatibility genes (*H*-1, *H*-2, etc.), because they control the transplantation of normal tissues as well as tumours. Most of the genes which have been identified so far have turned out to be allelomorphs at a few loci. At least one supposed locus, *H*-2, is probably a cluster of at least three closely linked loci (Hoecker, Counce & Smith, 1954; Amos, Gorer & Mikulska, 1955*b*). This pattern is similar to that of the genes which control the blood-group antigens in other mammals.

IV. THE ONE GENE-ONE ANTIGEN THEORY

On the basis of the genetical data which were available at the time, Haldane in 1933 proposed the one gene-one antigen theory. According to this theory, each of the genes governing tumour transplantation is responsible for the manufacture of a particular antigen, and for that antigen alone. The host only reacts to a transplanted tumour to destroy it, if the tumour possesses an antigen which is lacking in the host. Subsequent genetical experiments have all given results in accordance with the theory. So direct is the relationship between the genes and the antigens, that it has been suggested that the antigens must be primary gene products, or even identical with the genes themselves.

Apart from the purely genetical evidence, the theory has been supported by immunological experiments. Thus no transplantation antigen has been detected which is tissue-specific (with the exception of antigens of erythrocytes, and possibly chemically induced sarcomata), so that the genes appear to act independently of their cellular environment as well as of one another. For example, the antigenic stimulus from whole blood, spleen, or thymus, is able to induce heightened resistance against subsequent transplants of skin or tumour (Woglom, 1929; Billingham, Brent & Mitchison, 1957). These experiments on immunity show that the tissues must contain antigens in common, but a more radical conclusion can be drawn from experiments on actively acquired tolerance. Billingham, Brent & Medawar (1956*a*) have shown that leucocytes can induce complete tolerance to subsequent skin grafts, from which it follows that every antigen in skin is also represented in the leucocyte. A test for hybrid antigens on the same principle has been proposed by Haldane (1956), but not carried out.

It would also be expected that tissue which contains a double dose of genes should be more antigenic than tissue which contains only a single dose. This prediction has been tested by comparing the survival time of skin taken from inbred strains and F1 hybrids, but the available results are inconclusive (Prehn & Main, 1954; Billingham, Brent, Medawar & Sparrow, 1954; Barnes & Krohn, 1957).

The most striking confirmation of the theory has come from the recent work of Billingham, Brent & Medawar (1956b), in fractionating transplantation antigens from tissue. The antigens were found to be confined wholly to the nuclei, and were probably deoxyribonucleoproteins.

V. VARIATION AMONG TUMOURS

The variation which is found between tumours serves as a guide to the variation to be expected within single populations of tumour cells. Each tumour has a characteristic antigenic constitution, and differences are found between tumours which originate in the same inbred strain, and even in the same individual. A striking example of this variation is provided by three tumours which Cloudman (1932) obtained from a single inbred mouse. These tumours were tested by transplantation into hosts of an F2 cross with a foreign strain, where all three required two genes for their growth. The overlap of the genes was then tested by doubly inoculating the F2 mice with combinations of tumours. One tumour could not be tested satisfactorily; the other two required one gene in common, and one different.

The number of different transplantation antigens is much larger in normal tissue than in tumours. Barnes & Krohn (1957) found that every one of 120 grafts from A strain donors, and 154 from CBA donors, were eventually rejected by hosts of the F2 cross between the strains. It seems that the antigens which are represented in a tumour are a narrow and arbitrary selection of those which were present in the cell from which the tumour originated. Similar variation within clones of tumour cells is therefore to be expected, by an extrapolation of Darwinian principles.

In the experiment of Barnes & Krohn, a few skin grafts survived for as long as 100 days before being sloughed off. If transplants of a reasonably rapidly growing tumour with the same antigenicity had been made, these hosts would certainly have been killed. It follows, as Billingham, *et al.* (1954) have pointed out, that the transplantation of normal tissue is a more delicate test of antigenic difference than is tumour transplantation.

VI. METHODS OF IDENTIFYING
TRANSPLANTATION ANTIGENS

In most of the genetical experiments, antigens were only recognized if they could cause the rejection of a tumour transplant. This test may fail to reveal some of the weaker antigens in tumours, as comparison with skin grafting has shown. An antigen is weak in this sense only because it does not cause the rejection of the transplant; an antigen which appears to be weak in a fast-growing tumour may in fact provoke as much response as an apparently strong antigen in a slow tumour. Weak antigens were first demonstrated by Gorer (1947), who showed that tumour 15091a could grow and kill mice in the presence of very high titres of haemagglutinating antibody. The experiment shows that the tumour contained an antigen, which could elicit the production of antibody, but could not cause rejection of the transplant.

Gorer's experiment is an example of the use of a serological reaction to detect a weak antigen. Various refinements have since been added to this method. The capacity of tumour cells to absorb antibody, as well as elicit antibody production, has been used by Amos (1956); and Klein, Klein & Révész (1957) have carried out analogous tests using a titration of the cytotoxic activity of serum devised by Gorer & O'Gorman (1956). The tests based on serological reactions have the disadvantage that they are an indirect way of detecting the antigens responsible for transplantation immunity. These antigens may not be identical with those which participate in serological reactions. They stand to one another, according to a hypothesis proposed by Billingham et al. (1956b), in the relationship of primary and secondary gene products. But the inference from a negative serological reaction is probably safe, for no case is known of a cell containing a transplantation antigen without the corresponding agglutinating antigen.

Weak tumour antigens can be detected in a more direct way, by tests based on transplantation immunity. The tests can be made in two quite different ways, according to whether the tumour is used to provoke, or react with a state of immunity. According to the first method, a tumour which is suspected of possessing a weak antigen is transplanted into a host which lacks the antigen, but which would normally succumb to the tumour. After any antigen from the tumour has had time to reach the lymph nodes of the host, the tumour is killed or removed. The host is then tested for heightened resistance against a transplant of tissue which is known to possess the antigen in question. This test has been used by Feldman & Sachs (1957), who washed the peritoneum out with saline in order to remove ascites cells of the first tumour transplant.

The second method is to transplant the tumour under conditions which tip the balance in favour of the host. This can be done by pre-immunizing the host by prior transplantation, or alternatively by weakening the cells of the tumour. Pre-immunization was used by Amos, Gorer & Mikulska (1955 a) to reveal four weak antigens of tumour E.L. 4, in addition to the one which was normally revealed in a genetic test.

These tests can be used to identify, as well as to detect, weak antigens. For example, Hoecker, Counce & Smith (1954) described the transplantation of a C3H (H-2^k/H-2^k) tumour into the B10.D/2 strain (H-2^d/H-2^d), which is isogenic with C57BL (H-2^b/H-2^b). Prior immunization of the hosts with C57BL tissue accelerated the breakdown of the transplants. It follows that the H-2^k allele must produce some antigen in common with H-2^b, absent from H-2^d. This transplantation antigen corresponds to antigen E, detectable by haemagglutination. This example shows that an adequate panel of isogenic strains could be used to identify the separate transplantation antigens of a tumour from any strain.

VII. MUTATIONS OF TUMOURS IN COMPATIBLE HOSTS

The normal practice is to maintain tumours for genetical experiments by passing through hosts of the strain of origin. Under these conditions the antigenic constitution of tumours remains constant over long periods, through large numbers of cell divisions. Even then, in the apparent absence of known selective influences, clear-cut mutations of the antigens have occasionally been recorded, and many tumours maintained by serial passage apparently undergo a mild, but progressive evolution.

Clear-cut mutations have been discovered when the same tumour has been tested in F2 and back-cross mice on different occasions. For example, Strong (1926) tested in two successive years the *dbr*D tumour of the DBA strain in F2 crosses with the A strain. When first tested, the tumour required seven genes for growth. When tested again, two sublines of the tumour each required two genes, and a third required only one gene. Changes of this sort seem to occur suddenly, and tend to affect several antigens at the same time. The changes have always been towards a reduction in the number of genes required for growth.

VIII. IMMUNOLOGICAL SELECTION

Several attempts have been made to select mutations in tumour cells, by maintaining large populations of cells for as long as possible under conditions where most of the cells are being killed by the homograft reaction.

The usual way of doing this is to transplant the tumour into a foreign strain, where it can be serially passaged provided the transplantations are made at short enough intervals. Tumours treated in this way often remain unchanged for long periods. The cells may be slightly damaged, so that their weak antigens can manifest themselves in genetical tests, but the damage is transient. For example, Hauschka, Kvedar, Grinnell & Amos (1956) transplanted a C3H tumour, 6C3HED, for eighty-four serial passages in ICR Swiss mice. At the end of this heroic experiment the tumour was still unable to grow progressively in the Swiss mice, and two or three antigens instead of the usual single one were revealed by transplantation into a back-cross. Occasionally a tumour loses strain specificity when maintained under these conditions, but whether or not this happens seems to depend on personal idiosyncrasies of the tumour. Sachs & Gallily (1956a) passaged ten tumours under the same conditions, and found only two which changed, of which one had been growing irregularly to start with.

Certain obstacles to immunological selection are inevitable in the standard inbred strains. A tumour from a standard strain must alter several gene requirements to become compatible with another standard strain, and the selective pressure across the intermediate stages may be slight. Nor can the intermediate stages be detected without elaborate genetical tests. These difficulties do not apply to isogenic strains, which were originally designed with the possibility of mutation studies in mind. Using these strains, a test can be designed to detect a single mutation in a single tumour cell.

Several tumours from inbred strains have been repeatedly transplanted into the corresponding isogenic hosts. No case of progressive growth has been observed, even when the tumour cells were irradiated before transplantation (Mitchison, 1956b and unpublished; Klein & Klein, 1956). This failure is probably due to the use of homozygous tumour cells, which required a double mutation (or other complex genetic process) to become compatible with their hosts. The result might have been expected, because the breeding of isogenic strains depends on the fact that tumours do not often mutate in this way. This stability of the isogenic strain tumours allowed the Kleins to carry out a model experiment to confirm the assumption that a few compatible cells can survive while surrounded by similar incompatible cells which are being destroyed. Mixed inocula were prepared from compatible and incompatible tumour cells, which gave successful growth even with as small a proportion as 4×10^{-7} of compatible cells.

Heterozygous cells are more suitable for detecting mutations than homozygous ones, as has been realized independently several times (Lederberg, 1956; Mitchison, 1956b; Klein & Klein, 1956). A tumour from an F1 hybrid between two of Snell's isogenic strains has the special advantage that it should be heterozygous at a single histocompatibility locus. One of these tumours would normally be unable to grow progressively in either of the parental strains, but a single mutation (or other simple genetic process) would be sufficient to make the tumour compatible. This hypothesis has now been tested independently twice, once with an H-2^b/H-2^d tumour by Mitchison (1956 b), and once with H-2^a/H-2^s tumours by Klein *et al.* (1957).

A striking feature of these experiments is how differently individual tumours behave, even when they are histologically similar and have been induced by the same carcinogen. Three of the H-2^a/H-2^s sarcomata would grow in neither parental strain, one grew non-specifically in both and other strains, and one grew in a fraction of one parental strain; while the H-2^b/H-2^d carcinoma grew in a fraction of both parental strains. The last two of these tumours gave evidence of mutation and immunological selection. The growth of some transplants in the parental strains was preceded by a long latent period, which suggested that most of the initial cell population was killed. The tumours which were obtained from successful transplants in the parental strains proved to be variants with an altered host range. These variants were altered genetically, for they were stable after passage through F1 hosts. While one variant grew non-specifically, others were specific for the host in which the variant had originated. Several tests indicated that the specific variants had lost only their incompatible antigens. Thus a variant of the H-2^b/H-2^d tumour obtained from one parental strain would kill all hosts of this strain, while its behaviour in the other parental strain was unaltered. A careful search could detect no trace of the H-2^a antigen in variants of the H-2^a/H-2^s tumour which had been grown in H-2^s/H-2^s hosts. The variants grew equally well in normal and pre-immunized H-2^s/H-2^s hosts, and serological tests for the H-2^a antigen in the tumour cells were entirely negative.

The somatic mutation rate cannot be calculated from the available data without making several untested assumptions. For example, the number of divisions which the tumour cells went through before the homograft reaction commenced is not known. This multiplication was probably a significant factor in increasing the chances of a transplant surviving, for the frequency of progressive growth of the H-2^b/H-2^d tumour in the parental strains dropped to zero in pre-immunized hosts (Mitchison, unpublished data). Nor is it known whether the separate variant growths were all due to a single clone of variant cells in the original tumours. Irradiation of the H-2^a/H-2^s tumour increased the frequency of variant growths, as might be expected if each variant represented a separate mutation.

IX. THEORIES OF MUTATION

The mechanism of mutation in tumour cells has long been debated. The experimental facts which any explanation would have to account for are clear. Tumours differ from one another in their host range, or more precisely, in the genes they require in their hosts; and differences of the same sort are found among the cells of a single tumour. The immunological basis of these changes has to be settled, and how they are controlled by the genes.

The explanations which have been proposed fall into three main groups. The oldest is the somatic-mutation hypothesis of Little and Strong. According to this hypothesis, mutations occur in somatic cells in the same way as they occur in germinal cells. If one of the genes controlling an antigen mutates in a tumour cell, the antigen is altered or lost (the original proponents of the theory referred to transplantation factors or gene requirements, instead of antigens). A change in host range is thus brought about by all-or-nothing changes in antigens. This is a simple mechanism which has appealed strongly to geneticists. Some support for it has been provided by the discovery of mutations in the germ line, affecting transplantation antigens. Unfortunately, the mutations have been found only in inbred strains, where their status is in doubt. With the exception of the recent experiments on heterozygous tumours, little further evidence has accumulated in favour of somatic mutations.

Some data from the early genetical experiments do not fit the hypothesis. Changes in antigens seemed to take place in blocks during the origin of tumours from normal tissue, as well as within transplanted tumours. These block changes cannot be caused by point mutation if the one gene-one antigen relationship holds true.

The most recent experiments tend to show that all-or-nothing changes in tumour antigens are the exception rather than the rule. The most common event seems to be a conversion of one or more antigens from the strong to the weak state. That is, from controlling the normal host range of the tumour, they change to the control of antigens which can be detected only by subtler tests. An example of this has already been quoted, the experiments of Amos, Gorer & Mikulska with tumour E.L. 4, and a number of similar cases have also been reported.

A third reason why somatic mutation will not do is that changes in host range seem to be often associated with changes in chromosome morphology. This observation counts against the hypothesis, because point mutations in general do not produce morphological effects on the chromosomes. Opinion about the exact relationship between host range and the chromosomes has been changing. At present, it seems that tumours with a wide transplantation range are invariably heteroploid, while the more specific tumours are diploid or occasionally tetraploid (Sachs & Gallily, 1956a; Hauschka et al. 1956). This relationship must be taken into account by any theory of mutation. One general implication is that the transplantability changes during carcinogenesis are qualitatively different from the final change to wide transplantability. For the second change, but not the first, is invariably accompanied by a change in chromosome ploidy.

The hypotheses of Gorer and Hauschka and their colleagues make up a second group. Changes in host range are still assumed to be the result of changes in the tumour cell antigens, but the changes are of a quantitative nature. Thus Gorer has suggested that a few antigens might be produced in abnormally large quantities, so that the others are crowded out from the cell surface. According to this view the weak antigens are those which are present in the tumour cells in a reduced amount. The changes in the rates at which different antigens are produced are due, according to Hauschka, to disturbances in the balance of gene action. Heteroploidy could cause these disturbances by changing the relative dosages of genes and by the position effect.

A third hypothesis has been proposed by Sachs and his colleagues. Changes in host range are here assumed to reflect increases in the ability of tumour cells to withstand immunological attack, and not changes in content of antigens. An antigen may then be present in exactly the same quantity in the cells of two tumours, but appear weak in one and strong in the other. Indeed it has been claimed that an antigen may seem weak because of over-production; a tumour may defend itself by secreting antigens which neutralize antibody outside the cells (Feldman & Sachs, 1957). The over-production is attributed to duplication of genetic material. Sachs & Gallily (1956a, b) have emphasized that a wide transplantation range is invariably accompanied by extra chromosomes, in addition to the normal diploid set.

The controversy over the interpretation of the cytological changes cannot be settled until more data are available. No series of repeated mutations of a single tumour has yet been analysed cytologically, and so we do not know whether identical changes in transplantation range are accompanied by the same pattern of chromosome change. The heterozygous tumours are promising material for this analysis.

The immunological issue is also in doubt. Several attempts have been made to find out whether the weak antigens of tumours are really present in reduced amounts per cell. Some have been based on serological methods (Amos, 1956), which suffer from the objection of indirectness which has already been mentioned; the objection applies even more strongly to quantitative measurements of antigens than to qualitative detection. Others have been based on direct measurements of transplantation immunity (Feldman & Sachs, 1957), but the measurements were made per unit transplant, and not per cell.

Resistance to antibody as the sole factor in determining host range is a new idea, and it is interesting to consider how far it can be taken. It gives an excellent explanation of the fact that the strongest transplantation antigens—the H-2 series—are also the most stable in tumour cells: the level of resistance would have to be very high to overcome H-2 incompatibility. Most of the specific requirements of tumours for numbers of genes in their hosts could be accounted for in this way, simply by variation in the general level of resistance. However, some genetical observations present difficulties, such as the tumours which Cloudman showed to have incompletely overlapping gene requirements in their hosts. To explain these, a mechanism is required which affects specific antigens.

The only cases where the experimental evidence strongly favours the complete loss of a single antigen are the mutations in heterozygous tumours. The most likely genetical explanations of these losses are deletion, somatic crossing over, or point mutation. Serological investigation showed that in at least one mutation the antigens D and K were lost simultaneously. Cross-overs have been reported between D and K, so this evidence counts against the point-mutation hypothesis.

The controversies over details of the mechanism of immunological change in tumour cells may in the long run turn out to be of secondary interest. In the meantime, the mutations provide a valuable tool for analysing the relationship between events in the chromosomes and other parts of the cell, and the control of the synthesis of specific antigens. They provide also a means of investigating differences between the characteristics of individual tumours. They may be useful as material for the study of mutagenesis (Dr C. Auerbach, personal communication).

X. POSSIBLE EXAMPLES OF RECOMBINATION

One of the most exciting ways of using the transplantation antigens is as markers in the search for genetic recombination. Recombination, if it occurs at all between tissue cells, should be detectable by the general methods which have been established for micro-organisms. These entail mixing two types of cells with at least two markers each; the mixed population is then exposed to an agent which selects cells which carry a combination of markers from the two original types. The usual difficulty of this experiment is to distinguish recombination from mutation of the markers. An experiment of this sort is carried out whenever a tumour is passaged through foreign hosts: the markers on one side are the antigens of the tumour, and neoplastic cell growth; on the other they are the antigens of the host, and normal cell growth. The observed stability of tumour antigens under these conditions is proof that genetic recombination can only be an exceptional event.

Certain changes which take place in tumours growing in F1 cross hosts are possibly the result of genetic recombination between the cells of the tumour and the host. These changes, which were discovered by Barrett and Deringer, affect the transplantation range of tumours in F2 and back-cross hosts. They were of the type which would be expected if the cells from some tumours gained transplantation antigens from their hosts, and others lost them. The possibility of selecting rare tumour cells in F1 hosts has been critically excluded by application of the fluctuation test. Unfortunately, the most recent experiments show that these changes cannot be interpreted simply as a transfer of the genetic determinants of the antigens. Klein et al. (1957; with references to earlier papers), have found that an ability to tolerate antibody is gained by the tumour cells, rather than antigens.

Another possible example of genetic recombination is provided by the transformation of tumours by passage through foreign hosts with actively acquired tolerance (Koprowski, Theis & Love, 1956). The most spectacular transformation was obtained with a rat hepatoma; after serial passage through mouse foetuses and babies this tumour was finally able to grow in adult untreated mice. Similar transformations of mouse tumours were obtained in foreign mouse strains, including adapting the C3H tumour, 6C3HED, to grow in ICR Swiss mice. If these transformations were simply the result of selection of mutant cells, it is not clear why similar selection of 6C3HED did not occur during serial passage in adult Swiss mice in the experiment of Hauschka, Kvedar, Grinnell & Amos. The possibility of selection has been further discussed by Medawar & Bawden (see after Koprowski et al. op. cit.).

REFERENCES

Amos, D. B. (1956). *Ann. N.Y. Acad. Sci.* **63**, 706.
Amos, D. B., Gorer, P. A. & Mikulska, Z. B. (1955a). *Brit. J. Cancer*, **9**, 209.
Amos, D. B., Gorer, P. A. & Mikulska, Z. B. (1955b). *Proc. Roy. Soc.* B, **144**, 369.
Andreini, P., Drasher, M. L. & Mitchison, N. A. (1955). *J. Exp. Med.* **102**, 199.
Barnes, A. D. & Krohn, P. L. (1957). *Proc. Roy. Soc.* B, **146**, 505.
Billingham, R. E. & Medawar, P. B. (1950). *J. Anat.* **84**, 50.

BILLINGHAM, R. E., BRENT, L. & MEDAWAR, P. B. (1956a). *Phil. Trans.* **239**, 357.

BILLINGHAM, R. E., BRENT, L. & MEDAWAR, P. B. (1956b). *Nature, Lond.* **178**, 514.

BILLINGHAM, R. E., BRENT, L., MEDAWAR, P. B. & SPARROW, E. M. (1954). *Proc. Roy. Soc.* B, **143**, 43.

BILLINGHAM, R. E., BRENT, L. & MITCHISON, N. A. (1957). *Brit. J. Exp. Path.* **38**, 467.

BRIGGS, R. & KING, T. J. (1956). *Cold Spr. Harb. Symp. Quant. Biol.* (in the Press).

CLOUDMAN, A. M. (1932). *Amer. J. Cancer*, **16**, 568.

EPHRUSSI, B. (1956). In *Enzymes: Units of Biological Structure and Function.* Ed. Gaebler, O. H. New York: Academic Press.

FELDMAN, M. & SACHS, L. (1957). *J. Nat. Cancer Inst.* **18**, 529.

FORD, C. E., HAMMERTON, J. L., BARNES, D. W. H. & LOUTIT, J. F. (1956). *Nature, Lond.* **177**, 452.

GORER, P. A. (1942). *J. Path. Bact.* **54**, 51.

GORER, P. A. (1947). *Cancer Res.* **7**, 634.

GORER, P. A. & MIKULSKA, Z. B. (1954). *Cancer Res.* **14**, 651.

GORER, P. A. & O'GORMAN, P. (1956). *Transplant. Bull.* **3**, 142.

HALDANE, J. B. S. (1933). *Nature, Lond.* **132**, 265.

HALDANE, J. B. S. (1956). *J. Genet.* **54**, 54.

HAUSCHKA, T. S., KVEDAR, B. J., GRINNELL, S. T. & AMOS, D. B. (1956). *Ann. N.Y. Acad. Sci.* **63**, 683.

HOECKER, G. F., COUNCE, S. J. & SMITH, P. (1954). *Proc. Nat. Acad. Sci., Wash.* **40**, 1040.

KALISS, N. & SNELL, G. D. (1951). *Cancer Res.* **11**, 122.

KLEIN, G. (1956). *Annu. Rev. Physiol.* **18**, 13.

KLEIN, E. & KLEIN, G. (1956). *Nature, Lond.* **178**, 1389.

KLEIN, E., KLEIN, G. & RÉVÉSZ, L. (1957). *J. Nat. Cancer Inst.* **19**, 95.

KOPROWSKI, H., THEIS, G. & LOVE, R. (1956). *Proc. Roy. Soc.* B, **146**, 37.

LAW, L. W. (1954). *Advanc. Cancer Res.* **2**, 281.

LEDERBERG, L. (1956). *Ann. N.Y. Acad. Sci.* **63**, 662.

LINDSLEY, D. L., ODELL, T. T. & TAUSCHE, F. G. (1955). *Proc. Soc. Exp. Biol., N.Y.* **90**, 512.

MEDAWAR, P. B. (1947). *Biol. Rev.* **22**, 360.

MITCHISON, N. A. (1955). *J. Exp. Med.* **102**, 157.

MITCHISON, N. A. (1956a). *Brit. J. Exp. Path.* **37**, 239.

MITCHISON, N. A. (1956b). *Proc. R. Phys. Soc. Edinb.* **25**, 45.

MITCHISON, N. A. (1957). *J. Cell. Comp. Physiol.* (in the Press).

MITCHISON, N. A. & DUBE, O. L. (1955). *J. Exp. Med.* **102**, 179.

PREHN, R. T. & MAIN, J. M. (1954). *J. Nat. Cancer Inst.* **15**, 191.

PUCK, T. T. & MARCUS, P. I. (1955). *Proc. Nat. Acad. Sci., Wash.* **41**, 432.

SACHS, L. & GALLILY, R. (1956a). *J. Nat. Cancer Inst.* **16**, 803.

SACHS, L. & GALLILY, R. (1956b). *J. Nat. Cancer Inst.* **16**, 1083.

SNELL, G. D. (1948). *J. Genet.* **49**, 87.

SONNEBORN, T. M. (1950). *The Harvey Lectures*, series 47. Springfield, Ill.: C. C. Thomas.

STRONG, L. C. (1926). *Genetics*, **11**, 294.

TRINKAUS, J. P. (1956). *Amer. Nat.* **90**, 273.

WOGLOM, W. H. (1929). *Cancer Rev.* **4**, 179.

L'ACTION INHIBITRICE DU GLYOXAL SUR LES MACROMOLÉCULES BIOLOGIQUES

Par J. ANDRÉ THOMAS

Professeur de Biologie cellulaire à la Sorbonne,
26 rue d'Ulm, Paris (5e)

I. INTRODUCTION

L'action virulicide du glyoxal vient d'être découverte, de façon indépendante et simultanément aux États-Unis (Tiffany et al. 1957; Moffett, Tiffany, Aspergren & Heinzelman, 1957; Wright, Lincoln & Heinzelman, 1957; Limans et al. 1957); en Hollande (de Bock, Brug & Valop, 1957) et en France (André Thomas & Hannoun, 1957 a, b).

Or, l'action inhibitrice du glyoxal parait générale: elle s'étend aux bactériophages (André Thomas, Barbu & Cocioba, 1957) et à d'autres virus que celui de la grippe, aux bactéries gram + et gram − (André Thomas & Hannoun, 1957 a) et aux chromosomes animaux et végétaux (André Thomas & Deysson, 1957). Cette action pose des problèmes théoriques et aussi des problèmes pratiques.

Cependant, il semble, en fin de compte, que l'intérêt des questions soulevées réside surtout dans la possibilité de découvrir une série de dérivés du glyoxal, ayant même action que ce corps mais dépourvus de toxicité, qui puissent permettre d'agir in vivo sur les virus et sur les chromosomes.

Plusieurs propriétés du glyoxal sont bien connues et utilisées couramment dans l'industrie. Ce corps O=CH—CH=O, premier terme des dialdéhydes aliphatiques, est plus stable que les autres représentants de cette série. Sa volatilité est très faible, il a une légère odeur douce, il n'irrite pratiquement pas la peau: il se présente donc comme étant très différent du formol, premier terme de la série des mono-aldéhydes aliphatiques. Le glyoxal commercial est généralement distribué en solution aqueuse à 30%, incolore ou à peine jaune, mélange de diverses formes hydratées de glyoxal et de petites quantités de produits apparentés, ne modifiant pas notablement, en général, les effets du glyoxal monomère, acides organiques, formaldéhyde.* Par suite de sa double fonction aldéhyde, le glyoxal contracte essentiellement des liaisons avec les groupes protéiques et polyhydroxyliques: il forme des ponts éthyléniques entre les groupes

---

* Les présentations du glyoxal sous forme de combinaisons bisulfitiques sont moins actives aux concentrations convenables.

amine ou amide, d'une part, et entre les groupes amylacés, cellulosiques, vinyliques, etc., d'autre part. Par contre, le formol, dans les mêmes conditions, établit des ponts méthyléniques plus courts. Les macromolécules synthétiques ou d'origine naturelle peuvent donc par l'introduction du glyoxal, acquérir des propriétés nouvelles, résistance mécanique, résistance à l'eau pouvant aller jusqu'à l'insolubilisation selon le traitement, etc. différentes de celles que peut leur conférer le formol.

Les voies de recherches suivies par les trois équipes sus-mentionnées, américaine, hollandaise, française, sont différentes.

Le point de départ des travailleurs américains a été que l'aldéhyde β-isopropoxy-α-cétobutyrique détruit le virus de la maladie de Newcastle *in ovo*; d'où la recherche très poussée dans 6 classes de corps, de structures chimiques connues ou nouvelles, nécessaires à une telle action. De nombreux glyoxals aromatiques se montrent très actifs vis-à-vis des virus de la maladie de Newcastle et de la grippe en culture *in ovo* sur embryon de poulet; mais ces corps sont pour la plupart insolubles dans l'eau ou trop toxiques. Les acétals de glyoxal sont inactifs. D'autre part, de nombreux dérivés de l'aldéhyde β-amino-lactique, contenant un groupe amine secondaire ou tertiaire sont actifs, en particulier les chlorhydrates de β-diéthylamino- et β-di-isopropyl-aminolactaldéhyde. Enfin, l'hydrate de la β-éthoxy-α-cétobutyraldéhyde ('kethoxal') qui est actif *in ovo*, a une action probablement virulicide directe sur les virus au stade extracellulaire.

De leur côté, les chercheurs hollandais sont partis du fait que des corps ayant une structure α, β-dicarbonylée, ont été signalés comme ayant une activité antivirale. Ils ont testé une telle action sur le virus de la grippe cultivé sur œufs embryonnés âgés de 11 jours. L'introduction du corps à essayer était faite dans le liquide allantoïdien une heure avant l'inoculation du virus; après 48 heures d'inoculation, le titre du virus était déterminé par hémagglutination. 34 corps ont été essayés, 4 dérivés du glyoxal seulement sont actifs.

Ici encore, il s'agit vraisemblablement d'une action virulicide directe. Le traitement préalable du virus par le glyoxal est inactivant, selon les conditions de concentration ou de durée.

Pour notre part, nous avons suivi une autre voie. Des recherches sur l'atténuation du virus de la fièvre aphteuse (André Thomas, 1956), nous ont conduit à étudier l'action antivirale de nombreux produits de condensation par le formol, d'acides aminés et de leurs dérivés. C'est ainsi que 254 préparations ont été essayés sur le virus de la grippe en culture sur œufs de poule embryonnés. Puis, à titre de comparaison, le formol a été remplacé par le glyoxal et enfin le glyoxal utilisé seul, comme témoin d'expérience. Nous avons alors constaté la grande activité virulicide propre du glyoxal.

Ces recherches ont été poursuivies concurremment sur le virus de la fièvre aphteuse *in vivo* (non publiés encore), sur le virus de la grippe et sur d'autres virus *in ovo*.

Plus préoccupés d'analyser l'action du glyoxal et de vérifier sa généralité, que de rechercher de suite un dérivé complexe d'application pratique, nous nous sommes d'abord limités à l'étude du glyoxal proprement dit, le premier et le plus simple terme de la série, et nous avons étendu nos recherches à d'autres virus, aux bactériophages, aux bactéries et aux chromosomes.

Il est préférable, pour les recherches expérimentales, de conserver le glyoxal en solution concentrée à 30%, de neutraliser, puis de diluer stérilement, seulement au moment de l'emploi. Toute intervention comme le chauffage ou la filtration bactériologique est inutile (à cause des propriétés bactéricides), ou contre-indiquée (diminution de l'activité).

II. ACTION DU GLYOXAL SUR LE VIRUS DE LA GRIPPE

(1) *Inhibition de la multiplication du virus en culture sur embryon de poulet*

Technique. L'inoculation des œufs de poule, au 11ème jour d'incubation, est pratiquée dans l'allantoïde avec 0·1 ml. d'une dilution de virus grippal A souche PR8, correspondant à 100 doses infectantes. La dose de glyoxal est injectée par la même voie, ¼ d'heure plus tard. Il convient de remarquer que le glyoxal peut être introduit sans dommage dans l'allantoïde, à des doses dépassant le milligramme. Chaque série d'expérience comprend en général 10 œufs témoins, uniquement inoculés, et des lots de 10 œufs inoculés et traités. L'incubation est poursuivie pendant 27 heures environ à 35° C.; les œufs sont refroidis à +4° C. pendant une nuit pour éviter les hémorragies, puis les liquides allantoïdiens de tous les embryons vivants de chaque lot sont prélevés: on titre alors leur pouvoir hémagglutinant et on calcule le nombre de doses hémagglutinantes produites par embryon dans chaque lot: le nombre de doses est exprimé en pourcentage par rapport à celui du lot témoin (100%). L'analyse statistique montre que les différences de plus de 6% par rapport aux témoins sont significatives (Hannoun, 1954).

(*a*) *Action du glyoxal selon la dose.* Le tableau 1 montre que l'inhibition est constante et totale à la dose de 500 μg. par œuf, elle est supérieure à 90% à 250 μg., elle est de l'ordre de 50% pour 100 μg. environ. Par contre, l'inhibition due à la formaldéhyde dans les mêmes conditions, est inférieure à 50%, à la dose de 250 μg.

(*b*) *Action du glyoxal selon le délai.* L'influence du délai entre l'inoculation du virus et l'injection du glyoxal a été analysée. Les embryons reçoivent une injection de 500 μg. de glyoxal, à divers intervalles de temps après l'inoculation du virus, puis le protocole expérimental précédemment décrit est appliqué (Tableau 2).

Nous constatons que le glyoxal inhibe la multiplication du virus de la grippe lorsqu'il est injecté peu après l'inoculation virulente, et de nouveau

Tableau 1

Glyoxal				Formaldéhyde		
Œufs par expérience	Embryons morts	Virus (témoin 100 %) (%)	Dose (μg.)	Œufs par expérience	Embryons morts	Virus (témoin 100 %) (%)
10	5	0	10000	—	—	—
10	0	0	1000	10	2	0
9	0	0	1000	10	1	0
—	—	—	750	10	0	0
9	0	0	500	9	1	37
9	0	0	500	10	0	12
9	0	0	500	—	—	—
10	0	0	500	—	—	—
10	1	3	500	—	—	—
10	0	2	500	—	—	—
9	0	6	250	9	0	59
10	4	0	250	9	0	62
8	0	6	250	—	—	—
10	2	4	250	—	—	—
10	0	9	250	—	—	—
9	1	7	250	—	—	—
10	1	11	250	—	—	—
10	0	6	250	—	—	—
9	0	13	250	—	—	—
8	0	5	200	9	0	85
10	0	4	200	—	—	—
10	0	12	200	—	—	—
9	0	28	150	9	0	95
8	1	23	150	—	—	—
10	0	8	150	—	—	—
10	0	21	150	—	—	—
10	0	26	150	—	—	—
9	1	17	125	—	—	—
10	2	0	125	—	—	—
10	0	36	125	—	—	—
9	0	62	100	10	1	115
8	0	55	100	—	—	—
8	0	10	100	—	—	—
10	1	61	100	—	—	—
9	0	100	10	—	—	—

à partir de 6 h. 30 après; entre ces périodes, existe une phase de sensibilité moindre, même pour la dose relativement forte de 500 μg. de glyoxal. Cette phase se situe approximativement entre la 3ᵉ et la 6ᵉ heure après l'inoculation. Or, ce délai correspond sensiblement à la période de multiplication intracellulaire des particules virales; au contraire, après 6 h., le virus du premier cycle de multiplication est libéré et l'injection de glyoxal redevient presque totalement inhibitrice. D'autres séries expérimentales que celles indiquées au Tableau 2 donnent des résultats de même signification; en outre, on peut constater, dans les heures suivantes, la reprise d'une phase de moindre sensibilité.

(2) *Pouvoir virulicide.* La source de virus grippal A, PR8 est une culture en liquide allantoïdien d'embryon de poulet. On mélange dans un tube 1 ml. de ce liquide allantoïdien virulent et 1 ml. d'eau distillée contenant la dose de glyoxal; celle-ci n'est pas ajoutée au tube témoin. Les tubes sont portés une heure au bain-marie à 37°. On titre alors le virus restant: les

Tableau 2

Délai entre l'inoculation de virus et l'injection de 500 μg. de glyoxal	Œufs par expérience	Embryons morts	Virus (%)
Témoin	10	1	100
0 h. 15	9	0	1·8
3 h.	8	1	12
5 h. 30	8	0	22
6 h. 30	8	0	2·8
7 h.	8	1	2·1
7 h. 30	8	0	1·7

Tableau 3

Expériences	Témoins	Glyoxal		Formaldéhyde	
		Dose (μg.)	Virus restant	Dose (μg.)	Virus restant
1	7·7	200	<1	200	<1
2	7·6	50	<1	50	4·8
3	7·5	10	<2	10	7·5
4	8·1	1	7·4	10	7·9
5	8·2	10	1·6	—	—
		5	3·2	—	—
		1	6·8	—	—

liquides sont dilués de 10^{-1} à 10^{-9} et on inocule chaque dilution à cinq embryons de poulet de 11 jours, dans l'allantoïde. Après 3 jours d'incubation, le pouvoir hémagglutinant du liquide allantoïdien de chaque œuf est testé; la dilution 50% infectante est calculée selon Reed & Muench. Les résultats de ces expériences sont présentés sous la forme du logarithme de la dilution 50% infectante (Tableau 3).

Le glyoxal a donc un pouvoir virulicide intense vis-à-vis du virus grippal, ce pouvoir est au moins 10 fois plus grand que celui de la formaldéhyde.

III. ACTION DU GLYOXAL SUR LES BACTÉRIOPHAGES

Technique: Nous avons expérimenté avec le phage ϕ174, actif sur *Shigella dyssenteriae* YoR et qui est de très petite taille (ordre de 15 mμ); le phage D_4 (40 mμ) actif sur *Salmonella enteritidis*; le phage λ (ordre de 50 mμ) actif sur *Escherichia coli* K-12 S et le phage PS_{28}, actif sur le streptocoque C_{44}. La dose de glyoxal répartie en liquide de Ringer est mélangée à volume égal, avec une suspension de phages. Le nombre de bactériophages dans les mélanges est de 2 à 3.10^5/ml. Après divers temps de contact à la température du laboratoire, les mélanges sont dilués cent fois en liquide de Ringer, puis la détermination du nombre des bactériophages actifs est déterminée selon la méthode de Gratia.

D'après les résultats obtenus (Tableau 4 et Fig. 1) nous remarquons d'abord une sensibilité inégale des bactériophages étudiés.

Tableau 4. *Pourcentage d'inactivation par le glyoxal et par le formol de deux bactériophages, au cours du temps*

	Temps de contact en minutes	Glyoxal (µg./ml.)				Formol (µg./ml.)			
		500	250	100	50	500	250	100	50
Phage φ174	15	97	73	57	—	—	100	—	—
	15	92	—	40	20	100	—	97	84
	60	100	97	92	—	—	100	—	—
	60	100	—	81	50	100	—	100	99
	240	100	100	99	—	—	100	—	—
	240	100	—	100	90	100	—	100	100
Phage D_4	15	25	—	0	0	42	—	0	0
	60	37	—	6	10	66	—	33	23
	240	87	—	32	55	99	—	54	49

Fig. 1 a, 1 b. Inactivation de deux bactériophages: φ174 et D_4, par le glycol (G), et par le formol (F). 1 a, Dose: 100 µg./ml.; 1 b, En indice: minutes de contact.

Le phage φ174 est beaucoup plus sensible à la fois au formol et au glyoxal que le phage D_4. Ceci est en accord avec le fait que le phage φ174 est généralement plus sensible que le phage D_4 à tous les facteurs inactivants essayés. Toutefois, nous constatons qu'en 15 minutes de contact avec 100 µg./ml. de formol, 97 % des phages φ174 sont inactivés, tandis que l'inactivation du phage D_4 est inférieure à 10 %. Avec le glyoxal, dans les mêmes conditions, l'inactivation du phage φ174 est égale ou inférieure à 50 %, alors qu'il n'y a encore pas d'action sur les phage D_4.

D'autres expériences effectuées selon le même protocole ont montré que le phage λ a une sensibilité comparable à celle du phage D_4, tandis que le phage PS_{28} a une sensibilité intermédiaire, comprise entre celle du phage D_4 et celle du phage $\phi174$.

La concentration limite d'activité est voisine de 50 μg./ml. A cette concentration, l'inactivation par le formol du phage $\phi174$ est totale entre 1 et 4 heures; par le glyoxal elle est de l'ordre de 50 à 90 % dans le même laps de temps. Remarquons que dans les suspensions de phages D_4, même après un temps d'action de 48 heures, il se trouve encore 2 % de survivants.

Ainsi, le formol est plus actif que le glyoxal sur les bactériophages étudiés, alors que le glyoxal l'est au moins 10 fois plus que le formol sur le virus de la grippe APR8.

IV. ACTION BACTÉRICIDE DU GLYOXAL

Le pouvoir bactéricide du glyoxal ou de quelques-uns de ses dérivés a été signalée dans de rares cas par quelques auteurs.

(1) Des disques de papier de 10 mm. de diamètre, imprégnés de solution de glyoxal, soit à 10%, soit à 1%, puis séchés à 37° C., sont placés à la surface d'une gélose nutritive, en boîte de Pétri, après ensemencement en nappe du germe étudié. Au bout de 24 heures d'incubation, la largeur de l'anneau d'inhibition de la culture autour du disque est mesurée. Cet anneau d'inhibition est le plus souvent important, sa largeur varie selon la concentration du glyoxal et la nature des bactéries utilisées (Tableau 5).

(2) Très suggestive est aussi la comparaison, selon la même méthode et dans une même boîte de Petri, du pouvoir inhibiteur du glyoxal et des antibiotiques les plus actifs sur les germes considérés (disques d'antibiotiques de l'Institut Pasteur) (Tableau 6). D'autres témoins faits avec de la formaldéhyde à 10 % ne provoquent aucune inhibition par suite de l'évaporation de ce corps, lors de la dessiccation des disques.

(3) Il ne s'agit là que d'un test d'activité bactériostatique, mais celle-ci résulte en réalité d'un pouvoir bactéricide, comme l'établit l'expérience suivante. Des cultures repiquées de la veille sont ensemencées en bouillon ordinaire (2 gouttes de semence pour 10 cm.³); à ces cultures, on ajoute soit le glyoxal, soit la formaldéhyde, de façon à obtenir des dilutions connues. L'observation de l'inhibition de la croissance, traduisant le pouvoir bactéricide, est faite après 48 h. d'incubation (Tableau 7).

Le glyoxal a donc une activité bactériostatique et bactéricide pour des germes soit Gram +, soit Gram −. Nous retrouvons une activité bactéricide à l'égard de *E. coli* et du *B. pyocyanique* de même ordre que celle signalée. Toutefois, la formaldéhyde est plus active que le glyoxal sur les bactéries Gram − examinées; sur des bactéries Gram + (*S. aureus*), l'activité parait de même ordre; certaines bacteria (*E. coli*, *S. typhi*) peuvent encore donner une sub-culture après traitement par le glyoxal à $2 \cdot 10^{-3}$.

Tableau 5

	Largeur de l'anneau d'inhibition (mm.)	
	Glyoxal à 10 %	Glyoxal à 1 %
Staphylococcus aureus	13·5	5·5
Vibrio comma	—	4
Streptococcus faecalis	6·5	2·5
Listeria monocytogenes	—	2
Escherichia coli	8·5	0
Klebsiella pneumoniae (Friedlander)	6·5	0
Pseudomonas aeruginosa (pyocyanique)	5	1

Tableau 6

		mm.
Escherichia coli	Glyoxal 10 %	9·5
	Streptomycine	10·5
Salmonella typhi	Glyoxal 10 %	13·5
	Chloramphénicol	10
Staphylococcus aureus	Glyoxal 10 %	15·5
	Pénicilline	14

Tableau 7

Limite de la concentration bactéricide

	Glyoxal	Formaldéhyde
S. aureus	10^{-3}	10^{-3}
S. typhi	$0 \cdot 5 \cdot 10^{-4}$	10^{-4}
P. aeruginosa	10^{-3}	10^{-4}
E. coli	$0 \cdot 5 \cdot 10^{-3}$	10^{-3}

V. ACTION CHROMATOCLASIQUE DU GLYOXAL

Technique: Les solutions de glyoxal portées à un pH voisin de la neutralité et diluées à l'eau distillée ont été soit injectées sous la peau du rat blanc mâle ou femelle, soit utilisées comme milieu d'immersion pour les racines de bulbes d'*Allium sativum* L. et de plantules de *Pisum sativum* L. Après un délai variable, les rats ont été sacrifiés et leurs principaux viscères soumis à l'examen histologique, tandis que les racines baignées à nouveau dans le liquide de Knop dilué au demi fournissaient des prélèvements échelonnés pour examen cytologique direct (Carnoy, orcéine acétique, réaction nucléale).

(1) *Chromosomes animaux*

Les rats ont reçu, pour un poids de 200 gr. environ, soit 0 cm.³ 5 soit 1 cm.³ (dose toxique) d'une solution de glyoxal à 10 %; ils ont été sacrifiés après 12, 23 et 36 heures. Les résultats sont comparables aux deux doses utilisées. Après 12 heures, d'assez nombreuses mitoses, dans l'intestin grêle (duodenum) sont déjà profondément altérées; ce sont essentiellement des stathmométaphases à chromosomes agglutinés, avec présence de chromosomes ou de fragments de chromosomes aberrants. On observe quelques anaphases à chromosomes agglutinés, avec un ou plusieurs ponts chromosomiques (Pl. 1, figs. 1, 2). Dans la suite (23 et 36 h.), ces mitoses dégénèrent. La même agglutination des chromosomes se retrouve dans d'autres cellules à multiplication active comme le testicule (Pl. 1, fig. 3). Dans la rate, les pycnoses sont déjà nombreuses après 12 heures; elles sont d'une extrême fréquence à 23 heures (Pl. 1, fig. 4); on les observe de même, plus discrètes et plus tardives, dans d'autres organes (foie, pancréas).

(2) *Chromosomes végétaux*

Chez l'*Allium sativum* (noyau réticulé riche en chromatine), une solution de glyoxal à 10^{-4} n'exerce aucune action sensible pendant les trois premiers jours. A 10^{-3} se manifeste une inhibition mitotique de type préprophasique qui provoque la disparition des mitoses entre 24 et 48 heures; quelques ponts chromosomiques apparaissent dans les figures mitotiques les plus tardives. A 10^{-2}, l'intoxication est rapidement intense.

Mais, à la concentration de 5.10^{-3}, une action chromatoclasique importante se manifeste. Après 2 heures d'immersion, alors que l'activité mitotique est normale, la fragmentation des chromosomes est rare. Toutefois, après retour en liquide de Knop dilué au demi, on observe dans les heures suivantes (entre 3 et 42 heures) de nombreux fragments, ainsi que la présence de chromosomes restés dans le champ fusorial. Les fragments sont encore très peu fréquents à la métaphase (ordre de 0·4% des métaphases), ils se voient le plus souvent à l'anaphase (ordre de 10% des anaphases); ils peuvent être doubles (fragments d'origine métaphasique: Pl. 2, figs. 4–6), ou simples (fragments anaphasiques: Pl. 2, figs. 1–3); dans un petit nombre de cas, l'atteinte chromosomique se poursuit jusqu'à la pycnose (Pl. 2, fig. 7). Si l'immersion dans le glyoxal à 5.10^{-3} est prolongée au-delà de 2 heures (3, 6, 24 heures), les mitoses, encore nombreuses, comportent de plus en plus fréquemment des ponts chromosomiques; puis survient, à partir de 6 heures, l'agglutination des chromosomes métaphasiques, de même ordre que celle observée chez le rat. Après 48 heures, les cellules meurent, montrant de nombreuses mitoses abortives à chromosomes agglutinés et peu colorables.

Chez le pois (noyau réticulé pauvre en chromatine), les phénomènes sont analogues à ceux observés chez l'ail, mais la fragmentation des chromosomes est rare; les altérations consistent surtout en ponts chromosomiques ana- et télophasiques.

Signalons que des oxalates solubles (de sodium, d'ammonium) ne sont pas chromatoclasiques jusqu'à la concentration de 10^{-2}; mais celle-ci, après 3 heures d'immersion, entraîne la mort avec mitoses abortives à chromosomes agglutinés. A 10^{-3} pendant plus de 24 heures, une légère action mitoclasique se manifeste (rares stathmométaphases et cellules binucléées).

Le glyoxal, qui est virulicide et bactéricide, a donc une action qui s'exerce principalement sur les chromosomes. C'est une action chromatoclasique ou radiomimétique qui est de même nature, chez des cellules animales et végétales différenciées, en mitose; chez des cellules animales très sensibles, comme celles de la rate, il provoque de très abondantes pycnoses.

VI. DISCUSSION

(1) Les résultats obtenus montrent, d'une part que le glyoxal a une activité virulicide à l'égard d'un virus comme celui de la grippe, déjà intense à la concentration de 5 μg. par ml.; il s'agit principalement d'une action directe sur le virus au stade extracellulaire. Cette action est au moins 10 fois supérieure à celle du formol dans les mêmes conditions. Par contre, le formol, d'une manière générale, est à peu près deux fois plus actif que le glyoxal sur les bactériophages étudiés.

Parmi ceux-ci, le phage $\phi 174$ est de beaucoup le plus sensible: or, sa sensibilité au formol est du même ordre de grandeur que celle du virus de la grippe. Dans ces deux cas, nous remarquons que le glyoxal est très actif sur une structure virale essentiellement ribonucléoprotéique, et moins actif sur des bactériophages, essentiellement désoxyribonucléoprotéiques.

(2) On connaît les réactions du formol avec les fonctions amines libres des protéines. En outre, Fraenkel-Conrat (1954) a montré que le formol, à des concentrations qui sont 200 fois supérieures à celles que nous trouvons nécessaires pour inactiver les phages, modifie le spectre d'absorption des acides ribonucléiques, mais non celui des acides désoxyribonucléiques. Il conclut que l'inactivation des systèmes biologiques contenant de l'acide ribonucléique peut être due à ce nouveau mode d'action du formol. Nous avons donc comparé l'action du glyoxal et du formol sur les acides nucléiques.

Action sur les acides nucléiques

(*a*) En présence de 1 mg./ml. de formol ou de glyoxal, le spectre d'absorption ultraviolet d'une solution d'acide désoxyribonucléique de thymus de veau (45 μg./ml.) n'est pour ainsi dire pas modifié au cours du temps (Tableau 8).

Il n'y a pas non plus modification de la viscosité. De même, le spectre d'absorption ultraviolet d'une suspension de bactériophages D_4, contenant 5.10^{11} phages par ml., reste pratiquement inchangé dans les mêmes conditions.

(b) Par contre, sur l'acide ribonucléique de levure (40μg./ml.), le glyoxal et le formol (1 mg./ml.) ont une action comparable sur le spectre d'absorption, mais assez faible (Tableau 9 et Fig. 2).

Tableau 8. *Absorption dans l'ultraviolet (densité optique multipliée par 10^3) d'un mélange d'acide désoxyribonucléique de thymus de veau (45μg./ml.) et de formol ou de glyoxal (1000μg./ml.)*

	Temps	Longueurs d'ondes en mμ					
		240	250	260	280	290	300
Formol	0	675	960	1015	560	226	38
	3 h.	670	955	1015	563	240	42
	20 h.	660	945	1010	540	207	32
Glyoxal	0	741	995	1060	580	246	57
	3 h.	750	1000	1060	595	270	68
	20 h.	735	980	1040	570	234	53

Tableau 9. *Absorption dans l'ultraviolet (densité optique multipliée par 10^3). Mélange a $37°$ C. d'acide ribonucléique de levure (40μg./ml.) et de formol ou de glyoxal (1000μg./ml.)*

	Temps	Longueurs d'ondes en mμ							
		240	250	260	265	270	280	290	300
Formol	0	725	980	1030	955	825	512	186	36
	1 h.	740		1080			548		42
	4 h.	740		1100			564		
	24 h.	710	1015	1100		910	570	225	45
Glyoxal	0	810	1025	1040	948	815	520	223	65
	1 h.	850		1080			540		83
	4 h.	870		1085			545		
	24 h.	860	1105	1080		840	542	255	80

Le fait que des concentrations de formol et de glyoxal 200 à 2000 fois inférieures à celles nécessaires à modifier le spectre d'absorption des acides ribonucléiques suffisent pour inactiver les bactériophages ou le virus de la grippe, suggère que ces corps n'agissent pas sur les acides nucléiques, au moins de façon directe, mécanisme que Fraenkel-Conrat admet comme le plus probable dans le cas de l'acide ribonucléique. Au contraire, l'inactivation des virus désoxyribonucléiques sans qu'il y ait modification corrélative de leur spectre d'absorption, indiquerait plutôt un autre mode d'action que celui d'un effet direct sur les acides nucléiques.

(3) Cependant, le glyoxal a une activité bactéricide qui, dans certains cas, semble équivalente à celle du formol; enfin, il manifeste une action

chromatoclasique importante sur les chromosomes animaux et végétaux des cellules en mitoses; il provoque intensément la pycnose des noyaux des cellules animales sensibles. Il s'agit donc là d'une action intracellulaire progressive qui finalement se traduit par une atteinte des structures désoxyribonucléiques.

L'action la plus certaine de corps comme le formol et le glyoxal est évidemment celle sur les groupes aminés des protéines. En effet, l'intégrité de ces groupes sur les bactériophages, par exemple, est précisément

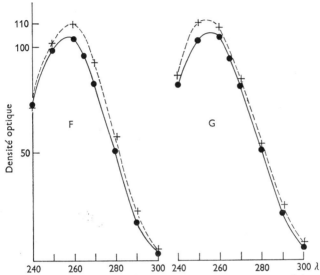

Fig. 2. Absorption dans l'ultraviolet d'un mélange d'acide ribonucléique de levure (40 μg./ml.) et de formol (F) ou de glyoxal (G: 1000 μg./ml.). Densité optique × 10³. Trait plein: valeurs initiales; tirets: après 24 heures à 37° C.

nécessaire pour la fixation de ceux-ci aux récepteurs bactériens (Puck & Tolmach, 1954). Il se peut qu'un mécanisme analogue entre en jeu dans le cas d'autres virus. Dans celui d'une action du glyoxal sur les désoxyribonucléoprotéines intracellulaires, le mécanisme réside vraisemblablement surtout dans la modification primaire des protéines associées à l'acide désoxyribonucléique. Cette modification parait retentir progressivement sur toute la structure, ce qui peut se traduire finalement, sur le plan cytologique, par la rupture l'agglutination et la dégénérescence des chromosomes. Bien entendu, le détail du mécanisme d'action sur les chromosomes est complexe; on peut invoquer plusieurs effets, qui d'ailleurs doivent s'associer: établissement de ponts éthyléniques, modification des liaisons de l'eau, d'ou influence sur la spiralisation, sur la matrix, augmentation de la rigidité, etc. Il se peut aussi que les constituants ribonucléiques du cytoplasme soient altérés.

La suggestion que l'action du glyoxal porte d'abord sur les protéines n'exclut pas, évidemment, la possibilité que d'autres processus entrent simultanément en jeu; mais l'action sur les protéines semble, jusqu'à plus ample informé, prépondérante.* Que l'effet du glyoxal soit pratiquement nul sur le spectre d'absorption et sur la viscosité de l'acide désoxyribonucléique, que cet effet soit assez faible sur le spectre de l'acide ribonucléique, il ne s'en suit pas que ce corps ne puisse quand même pas avoir un certain retentissement sur les fonctions physiologiques de ces acides. Il serait intéressant de tester l'action du glyoxal sur l'acide désoxyribonucléique bactérien transformant et de vérifier s'il n'a pas une influence génétique, alors que les constantes physiques de cet acide restent à peu près inchangées. D'une façon plus générales il parait indiqué de rechercher les effets mutagènes éventuels du glyoxal.

Cependant, une des difficultés d'utilisation est que chez l'animal, le glyoxal est oxydé notamment en acide oxalique, ce qui est dangereux. Remarquons, que d'après nos essais, ce n'est pas par l'action des oxalates solubles que s'explique l'effet biologique du glyoxal. Ce n'est pas non plus par la transformation dans l'organisme en d'autres produits d'oxydation, comme par exemple l'acide rhodizonique. Nous avons vérifié que les rhodizonates de sodium et de potassium ne sont pas virulicides (Thomas & Hannoun, 1957). Toutefois, nos travaux actuels montrent que la concentration du glyoxal nécessaire à l'action virulicide *in vivo* chez les Mammifères (concentration thérapeutique: ordre de $5 \cdot 10^{-4}$) est toxique. Ce n'est donc pas le glyoxal lui-même qui pourrait servir dans la lutte chimique, *in vivo*, contre certains virus, mais il doit être considéré comme un bon point de départ.

En effet, l'étude des propriétés inhibitrices du glyoxal sur les macromolécules biologiques a été abordée comme type d'action d'un corps beaucoup moins toxique que le formol et susceptible de fournir de nombreux dérivés. Or, nous pouvons considérer, d'après nos recherches en cours, que certains de ces dérivés conservent une action, dans l'organisme, tout en étant peu toxiques. Il n'est pas impossible que parmi ces corps, il s'en trouve qui permettent d'agir efficacement *in vivo*, tôt ou tard, sur les virus et sur les chromosomes.

En somme, le glyoxal est donc moins toxique que le formol, tout en possédant, d'une part, une activité virulicide remarquable, principalement sur les virus ribonucléoprotéiques, et d'autre part, une activité chromatoclasique.

---

* Elle s'exerce par exemple de façon intéressante sur la substance fondamentale du tissu conjonctif qu'elle gélifie, provoquant un oedème important et surtout durable que nous étudions.

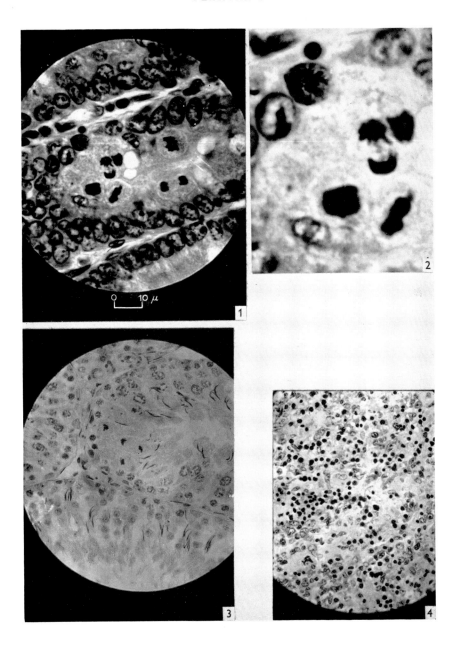

O 10 μ

1

2

3

4

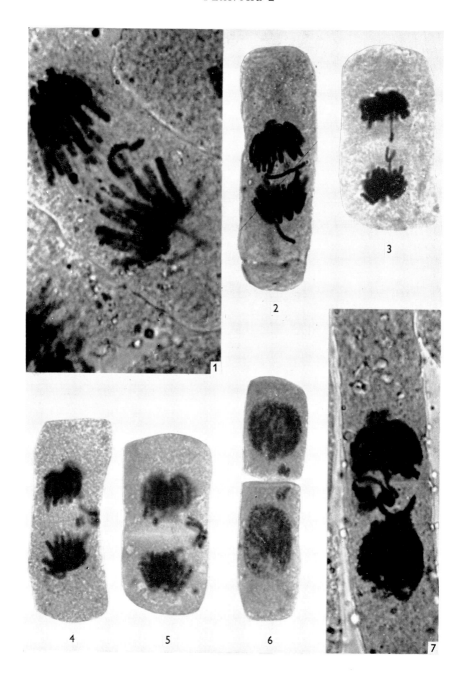

BIBLIOGRAPHIE

DE BOCK, C. A., BRUG, J. & VALOP, J. N. (1957). *Nature, Lond.* **179**, 706.

FRAENKEL-CONRAT, H. (1954). *Biochim. Biophys. Acta*, **15**, 307.

HANNOUN, CL. (1954). Thése Doct. ès Sciences, Université de Paris.

LIMANS, W. F., UNDERWOOD, G. E., SLATER, E. A., DAVIS, E. V. & SIEM, R. A. (1957). *J. Immunol.* **78**, 104.

MOFFETT, R. B., TIFFANY, B. D., ASPERGREN, B. D. & HEINZELMAN, R. V. (1957). *J. Amer. Chem. Soc.* **79**, 1687.

PUCK, T. T. & TOLMACH, L. J. (1954). *Arch. Biochem. Biophys.* **51**, 229.

THOMAS, J. ANDRÉ (1956). *C.R. Acad. Sci., Paris*, **242**, 694.

THOMAS, J. ANDRÉ, BARBU, E. & COCIOBA, I. (1957). *C.R. Acad. Sci., Paris*, **245**, 1182.

THOMAS, J. ANDRÉ & DEYSSON, G. (1957). *C.R. Acad. Sci., Paris*, **245**, 735.

THOMAS, J. ANDRÉ & HANNOUN, CL. (1957 *a*). *C.R. Acad. Sci., Paris*, **244**, 2258.

THOMAS, J. ANDRÉ & HANNOUN, CL. (1957 *b*). *C.R. Acad. Sci., Paris*, **244**, 2329.

TIFFANY, B. D., WRIGHT, J. B., MOFFETT, R. B., HEINZELMAN, R. V., STRUBE, R. B., ASPERGREN, B. D., LINCOLN, E. A. & WHITE, J. L. (1957). *J. Amer. Chem. Soc.* **79**, 1682.

TREFOUEL, J. (1957). *C.R. Acad. Sci., Paris*, **244**, 2262.

WRIGHT, J. B., LINCOLN, E. H. & HEINZELMAN, R. V. (1957). *J. Amer. Chem. Soc.* **79**, 1690.

EXPLICATION DES PLANCHES

PLANCHE 1

Figs. 1 et 2. Duodénum de rat ♀; glyoxal 10 %, 0·5 cm³/200 gr, sacrifice après 12 heures; *fix.* Limousin, *color.* hémalum, eosine-aurantia-orange.

Fig. 3. Testicule de rat; glyoxal 10 %, 0·5 cm³/200 gr, sacrifice après 23 heures; *fix.* Bouin, *color.* hémalum, éosine-aurantia-orange.

Fig. 4. Rate de rat ♂; glyoxal 10 %, 1 cm³/200 gr, sacrifice après 23 heures; *fix.* Bouin, *color.* hémalum, éosine-aurantia-orange.

PLANCHE 2

Racines *d'Allium sativum*: Immersion dans le glyoxal à 5·10⁻³ pendant 2 heures, puis dans le liquide de Knop dilué de moitié pendant 42 heures. *Color.* orcéine acétique.

Figs. 1, 2 et 3. Anaphases avec fragments chromosomiques simples.

Figs. 4, 5 et 6. Anaphases avec fragments chromosomiques doubles, d'origine métaphasique.

Fig. 7. Anaphase pycnotique.